Zoophysiology Volume 34

Springer
Berlin
Heidelberg
New York
Barcelona
Budapest
Hong Kong
London
Milan
Paris
Santa Clara
Singapore
Tokyo

Zoophysiology

Volumes already published in the series:

Volume 1: *P.J. Bentley*
Endocrines and Osmoregulation (1971)

Volume 2: *L. Irving*
Arctic Life of Birds and Mammals Including Man (1972)

Volume 3: *A.E. Needham*
The Significance of Zoochromes (1974)

Volume 4/5: *A.C. Neville*
Biology of the Arthropod Cuticle (1975)

Volume 6: K. *Schmidt-Koenig*
Migration and Homing in Animals (1975)

Volume 7: *E. Curio*
The Ethology of Predation (1976)

Volume 8: *W. Leuthold*
African Ungulates (1977)

Volume 9: *E.B. Edney*
Water Balance in Land Arthropods (1977)

Volume 10: *H.-U. Thiele*
Carabid Beetles in Their Environments (1977)

Volume 11: *M.H.A. Keenleyside*
Diversity and Adaptation in Fish Behaviour (1979)

Volume 12: *E. Skadhauge*
Osmoregulation in Birds (1981)

Volume 13: *S. Nilsson*
Autonomic Nerve Function in the Vertebrates (1983)

Volume 14: *A.D. Hasler*
Olfactory Imprinting and Homing in Salmon (1983)

Volume 15: *T. Mann*
Spermatophores (1984)

Volume 16: *P. Bouverot*
Adaption of Altitude-Hypoxia in Vertebrates (1985)

Volume 17: *R.J.F. Smith*
The Control of Fish Migration (1985)

Volume 18: *E. Gwinner*
Circannual Rhythms (1986)

Volume 19: *J. C. Rüegg*
Calcium and Muscle Activation (1986)

Volume 20: *J.-R. Truchot*
Comparative Aspects of Extracellular Acid-Base Balance (1987)

Volume 21: *A. Epple* and *J.E. Brinn*
The Comparative Physiology of the Pancreatic Islets (1987)

Volume 22: *W.H. Dantzler*
Comparative Physiology of the Vertebrate Kidney (1988)

Volume 23: *G.L. Kooyman*
Diverse Divers (1989)

Volume 24: *S.S. Guraya*
Ovariant Follicles in Reptiles and Birds (1989)

Volume 25: *G.D. Pollak* and *J.H. Casseday*
The Neural Basis of Echolocation in Bats (1989)

Volume 26: *G.A. Manley*
Peripheral Hearing Mechanisms in Reptiles and Birds (1989)

Volume 27: *U.M. Norberg*
Vertebrate Flight (1990)

Volume 28: *M. Nikinmaa*
Vertebrate Red Blood Cells (1990)

Volume 29: *B. Kramer*
Electrocommunication in Teleost Fishes (1990)

Volume 30: *W. Peters*
Peritrophic Membranes (1991)

Volume 31: *M.S. Kaulenas*
Insect Accessory Reproductive Structures (1992)

Volume 32: *A.L. Val* and *V.M.F. de Almeida-Val*
Fishes of the Amazon and their Environment (1995)

Volume 33: *R. Wiltschko* and *W. Wiltschko*
Magnetic Orientation in Animals (1995)

Volume 34: S. B. Moffett
Nervous System Regeneration in the Invertebrates (1996)

Stacia B. Moffett

Nervous System Regeneration in the Invertebrates

With 86 Figures

Springer

Dr. STACIA BRANDON MOFFETT
Washington State University
Department of Zoology
Pullman, WA 99164, USA

Library of Congress Cataloging-in-Publication Data

Moffett, Stacia B.
Nervous system regneration in the invertebrates / S.B. Moffett.
p. cm. -- (Zoophysiology ; v. 34.)
Includes bibliographical references and index.
ISBN-13:978-3-642-79841-2 e-ISBN-13:978-3-642-79839-9
DOI: 10.1007/978-3-642-79839-9

1. Nervous system--Regeneration. 2. Invertebrates--Nervous system. I. Title. II. Series.
QP363.5.M64 1996
592'.0188--dc20 95-41927
CIP

ISBN-13:978-3-642-79841-2

Cover design: Springer-Verlag, Design & Production
Typesetting: Mitterweger Werksatz GmbH, 68723 Plankstadt, Germany
Production: PROEDIT GmbH, 69126 Heidelberg, Germany
SPIN: 10055772 31/3137 – 5 4 3 2 1 0 – Printed on acid-free paper

Preface

Invertebrate animals represent a diversity of solutions to life's challenges. Success in a wide range of environments has been achieved by an almost bewildering range of invertebrate body forms. These body forms are reflected in the wonderful diversity of their nervous systems. Despite this apparent diversity, studies of the development of invertebrates and vertebrates are yielding common themes at the molecular level. Likewise, the phenomenon of neural regeneration is based upon properties intrinsic to neurons and responses to a remarkably conserved chemical language. This monograph focuses on the diversity and commonality of responses to neural injury.

The rough and tumble of life may frequently damage some part of the body, particularly the appendages or sensory systems. The nervous system is usually involved in repair of other body systems and often may itself require repair. Some animals are particularly successful in regenerating the nervous system or body parts. We particularly marvel at these feats of regeneration because we human beings are not particularly successful, despite our relatively long life and the advantages that would seem to accrue from such repair. It is no wonder that we would hope to learn the secrets of the more successful animals and strive to emulate them!

Mechanisms of neural regeneration are often more accessible in invertebrates than in vertebrates because questions of specificity are more easily addressed using the identifiable neurons of the relatively simpler nervous systems of some invertebrates. The shared molecular cues governing control mechanisms in invertebrate and vertebrate regeneration suggest that study of invertebrates may provide insights relevant to human neural repair.

Among the vertebrates and the invertebrates, there are great differences in the capacity for regeneration exhibited by a particular species. Recent work on a few species of invertebrates has revealed some of the principles whereby repair is achieved. These observations have also suggested novel approaches to promote neural repair. Intriguing observations on other invertebrates suggest an untapped reservoir for future studies.

Current progress in the study of nervous system regeneration, both that of vertebrates and invertebrates, is rapid. Individuals working with one invertebrate group are often more conversant with the relevance of their work to vertebrate studies than with related work on other invertebrates. My aim in this monograph was to bring together observations on many different kinds of animals so that similarities and differences could be appreciated. Many friends have offered encouragement and have generously shared insights and work in progress. Permission was generously granted by both the authors and the copyright holders for reproduction of the figures included with the text. I am especially indebted to Drs. J. S. Edwards, S. M. Fredman, R. L. Ridgway and D. F. Moffett for their comments on drafts of this monograph, and to Joan Folwell for technical assistance.

I hope that interest in the phenomenon of neural regeneration in invertebrates will continue to grow, and that this synthesis will spur new research, foster communication across this broad field, and bolster appreciation for the contributions that invertebrates can make to neural science.

Stacia Moffett

Contents

Chapter 1 **The Phenomenon of Neural Regeneration**

1.1 How Shall We Define Neural Regeneration? ... 1
1.1.1 What Systems Are Involved? ... 1
1.1.2 Intrinsic and Extrinsic Factors Compromise Repair ... 1
1.1.3 Neural Injury Reinitiates Growth ... 3
1.1.4 Injury Tests the Limits of Neural Plasticity ... 3
1.2 Replacement of Neurons ... 3
1.2.1 Where Will New Neurons Come From? ... 3
1.2.2 How Will New Neurons Find Their Way? ... 4
1.2.3 Which Animals Can Replace Neurons? ... 4
1.2.4 What Factors Delineate Neurogenesis? ... 6
1.3 Regrowth of Neurites ... 6
1.3.1 Incidence of Neurite Repair ... 6
1.3.2 Comparisons Between Development and Regeneration ... 7
1.4 A Cost-Benefit Analysis of Neural Regeneration ... 7
1.5 Rationale for Studying Invertebrate Regeneration ... 10
1.5.1 Invertebrates Offer the Simplest Systems ... 10
1.5.2 Invertebrates Offer a Great Diversity of Solutions ... 11
1.6 Conclusions ... 11

Chapter 2 **A Survey of Neural Repair in Invertebrates**

2.1 Introduction ... 13
2.2 Cnidaria: Neural Structure and Behavior ... 13
2.2.1 Examples of Regeneration ... 14
2.2.2 Regeneration in *Hydra* ... 14
2.3 Ctenophora: Comb Jellies ... 16
2.4 Platyhelminthes: The Simplest Bilateral Nervous System ... 16
2.4.1 Planarians: Masters of Body Regeneration ... 18
2.4.2 Role of the Nervous System in Body Repair ... 19
2.4.3 Repair of the Nervous System ... 20
2.5 Nemertea: Ribbon Worms ... 21

2.6 Nematoda: Roundworms 23
2.7 Annelida: Introduction to the Phylum 23
2.7.1 Polychaetes 24
2.7.2 Oligochaetes 25
2.7.3 Hirudinea 28
2.8 Arthropoda: Neural Organization and Repair 29
2.8.1 Reinnervation of Muscle 30
2.8.2 Refinement of Connectivity 32
2.8.3 Molting and Regeneration 32
2.8.4 Regeneration of Sensory Cells 33
2.9 Mollusca: Unsegmented Animals with a Range of Cephalization 34
2.9.1 Gastropods, The Snails and Slugs 34
2.9.2 Responses to Naturally Occurring Injuries 35
2.9.3 Axon Regeneration 38
2.9.4 CNS Regeneration 38
2.10 Echinodermata: The Spiny Deuterostomes 38
2.11 Chordata: Our Own Phylum 39

Chapter 3 **Early Responses to Neural Injury**
3.1 Introduction 41
3.2 Immediate Responses of Injured Neurons 41
3.2.1 Is the Injury Discharge a Cry for Help? 41
3.2.2 Basis of an Injury Discharge 42
3.2.3 Effects of Depolarization and Ca^{2+} Entry 42
3.2.4 Repairing the Axon Membrane 44
3.2.5 Changes in Excitability Following Axotomy ... 51
3.2.6 Excitability Changes: Was Injury the First Teacher? 53
3.3 Growth Following Axotomy 54
3.3.1 Initial Outgrowth is Independent of Soma Responses 54
3.3.2 Retrograde Signals 55
3.3.3 Anatomical Correlates of the Soma Response .. 56
3.3.4 Metabolic Adjustments Associated with Regeneration 58
3.3.5 Under What Conditions Is Regeneration Initiated? 59
3.3.6 Speculation on the Role of Diffusible Factors in Regeneration 59
3.4 Survival of Anucleate Axon Segments 61
3.4.1 Mechanisms of Distal Segment Survival 61
3.4.2 Consequences of Distal Segment Survival 61
3.4.2.1 Target Support by Surviving Distal Segments 61
3.4.2.2 Pathway Preservation 62

3.4.2.3 Proximal and Distal Segment Fusion . 62
3.4.2.4 Experimental Approaches to Axonal Fusion. 66
3.5 Responses of Nonneural Cells to Injury 67
3.6 What Activates Responses in Uninjured Neurons? . . 72
3.7 Conclusions 73

Chapter 4 **Pathfinding by the Growth Cone**
4.1 Introduction 75
4.2 Growth Cone Morphology 75
4.3 Growth Cone Extension. 78
4.3.1 Membrane Addition 78
4.3.2 Reshaping the Cytoskeleton 78
4.3.3 Responses to the Environment 80
4.4 Ion Channels in the Growth Cone 82
4.4.1 Recording Channel Activity 82
4.4.2 Dynamics of Growth Cone Channel Distribution . 83
4.4.3 Substrate Effects on Growth Cone Channels . . 84
4.4.4 Channel Expression Reflects Growth State. . . . 86
4.4.5 Electrical Synapses and Intercellular Communication 87
4.4.6 Developmental Regulation of Channel Expression. 88
4.4.7 Sensitivity to Electrical Fields 89
4.4.8 Roles of Electrical Activity in Growing Neurites . 89
4.4.9 Roles of Calcium 91
4.4.9.1 Calcium and the Cytoskeleton. 91
4.4.9.2 Sprouting and Collapsing 92
4.4.9.3 Ca^{2+} Effects Mediated by Second Messengers. 94
4.5 Responses of Growth Cones to Neurotransmitters 94
4.6 Differentiation Following Growth Cone Interactions 97
4.7 Receptors and Molecular Cues in the Environment 98
4.7.1 Functions of External Signals 98
4.7.2 Binding and Recognition Systems 99
4.7.2.1 Introduction 99
4.7.2.2 Homophilic Binding: Interactions between Like Molecules. 100
4.7.2.3 Heterophilic Binding: Interactions between Unlike Molecules. 101
4.7.2.4 Identification of Glycoproteins. 104
4.7.2.5 Antibodies, Blockers and Mutations Disrupt Pathfinding 106

4.7.2.6 Cues Expressed in Gradients 111
4.8 Conditioned Medium and Growth Promotors 111
4.8.1 Growth Requirements in Culture 111
4.8.1.1 Growth Factors in Gastropod Conditioned Medium 113
4.8.1.2 Neurosecretory Products as Growth Factors 116
4.8.1.3 Insulin-Like Molecules Support Sprouting 116
4.8.1.4 Inhibitors of Sprouting 117
4.8.2 Culture Influences on Branching Pattern 120
4.9 Designing Culture Conditions 120
4.10 Conclusions 123

Chapter 5 **Synapse Formation and Alteration During Regeneration**
5.1 Introduction 125
5.1.1 Comparing Synaptogenesis in Regeneration and Development 125
5.1.2 Novel Connectivity in Regenerated Nervous Systems 127
5.1.3 Reduced Plasticity in Regenerating Neurons... 128
5.1.4 Circuits Reformed in Vivo and in Vitro 128
5.2 The Road to Recovery 129
5.2.1 Behavior of Injured Animals 129
5.2.2 Connectivity Changes and Hypersensitivity... 131
5.2.3 Remodeling of Connections 132
5.2.4 Consequences of Deafferentation 135
5.2.4.1 Effects of Sensory Deprivation 135
5.2.4.2 Central Compensation for Sensory Deficits 139
5.2.5 Compensation for Ablated Ganglia 141
5.2.6 Ganglion Regeneration 143
5.3 How Neurons Select Targets 144
5.3.1 Limitation of Choices by Pathway Cues 144
5.3.2 Cell-Adhesion Compatibility in Synaptogenesis . 145
5.3.3 Specific Recognition Systems 148
5.3.3.1 Specificity in Arthropod Neuromuscular Innervation 148
5.3.3.2 Specific Connections Made by Transplanted Neurons 152
5.3.3.3 Mutants Elucidate Synaptic Specificity . 153
5.4 Development of a Functioning Synapse 154
5.4.1 Neurotransmitters in Synapse Formation 154
5.4.2 Targets May Induce Excitation-Secretion Coupling 155

5.4.3 Alterations in Ca^{2+} Channel Distribution 155
5.4.4 Electrical Activity in Synapse Formation 157
5.4.5 Responses to Growth Cone Contact with Target . 158
5.5 The Role of Competition........................ 160
5.6 Conclusions.................................... 161
References 163

Chapter 1

The Phenomenon of Neural Regeneration

1.1 How Shall We Define Neural Regeneration?

1.1.1 What Systems Are Involved?

Following neural injury, the complex sequence of events leading to neural regeneration will not be undertaken unless survival in the short term is assured. For most species to survive injury to the central nervous system (CNS), neural control of homeostatic functions must remain intact, although some very simple animals, such as flatworms, ribbon worms, or sea cucumbers, can sustain extreme injury and require only that enough neural tissue be present to provide a stimulus for regeneration. The nervous system apparently plays an organizing role in the regeneration of many animals (Chap. 2).

Although the nervous system is intimately involved in its own regeneration, the control and coordination exerted by nonneural systems can spell the success or failure of regeneration. Paracrine, hormonal, and immune signals, involving cells such as hemocytes and glia, contribute importantly to neural regrowth or block its progress (Chap. 3). Stimulation provided by the periphery may also play a role in repair of the CNS.

1.1.2 Intrinsic and Extrinsic Factors Compromise Repair

Repair that replaces all elements and connections produced during normal development clearly would qualify as regeneration. Rapid and specific repair of peripheral and central connections satisfies this definition, as does regrowth of an entire nervous system through regional recapitulation of development. Usually, however, errors are introduced in the repair process. The regenerating neurites may lose their way as they penetrate territory lacking adequate pathfinding cues. Even if they arrive at the correct target, they must compete with the responses of uninjured neurons to occupy vacated synaptic space.

In some cases, pathfinding is made easy by inherent features of the nervous system. Some general cues defining particular regions may remain marked in adult nervous systems. For instance, some songbirds add neurons episodically throughout life, necessitating retention of cues that direct neurite migration (Alvarez-Buylla and Nottebohm 1988). One might expect the pathfinding of

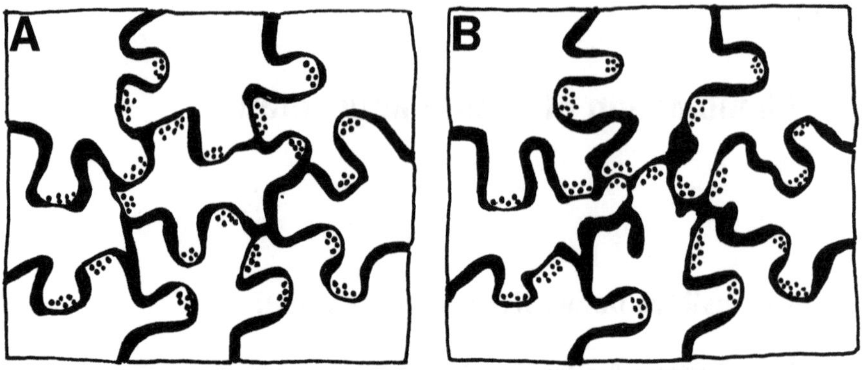

Fig. 1.1. The nervous system as a living jigsaw puzzle. **A** The puzzle piece in the center has been injured and will degenerate. **B** In the absence of the center puzzle piece, other pieces alter their shapes to fill the space

regenerating neurites to be successful in such nervous systems. Similarly, the mammalian olfactory system retains embryonic features, in keeping with its continuous generation of replacement sensory neurons (Graziadei and Monti-Graziadei 1980; Gonzalez and Silver 1994). In other animals which regenerate successfully, developmental cues disappear in the mature animal but are reexpressed during regeneration (Stuermer et al. 1992). Alternatively, regrowth may depend upon entirely different cues from those employed during development (Denburg and Whitington 1978; Fekete and Brockes 1987; Nardi and Vernon 1990).

If the mature nervous system is likened to a completed jigsaw puzzle, with all its interlocking pieces, then regeneration represents the process of trying to repair such a jigsaw puzzle, when some of the original pieces are removed (Fig. 1.1). However, in this living puzzle, the remaining pieces immediately begin to melt into one another and reach out to cover the gaps. Even in the uninjured state, the living jigsaw puzzle responds to use with subtle changes in the shapes of the pieces. The capacity for such structural and functional changes is greatly increased by injury, as described below.

When new neurons or entire ganglia are replaced through the partial recapitulation of development, newly generated repair elements must be integrated with intact regions of the nervous system to reestablish function. In even the most successful regeneration of connections, the resulting circuitry differs in subtle ways from the original condition. The challenge posed by extensive neural injury is not just to replace correct connections but to create substitute connections that satisfy functional demands.

1.1.3 Neural Injury Reinitiates Growth

The surviving neurons in an injured nervous system can be divided into injured cells, uninjured cells that respond to the injury, and the nonresponders. Responses of the uninjured neurons may include withdrawal of synapses (Perry et al. 1993) and reactive synaptogenesis (Rotshenker 1979; Brown et al. 1981; Blackshaw et al. 1982). Reactive synaptogenesis, the sprouting of new connections prior to or in the absence of replacement or regrowth of lost connections, can sometimes block reformation of the original synaptic contacts. However, when these novel connections are appropriate, i.e., they result in functional behavior, they have the potential to compensate for lost connections. Although reactive synaptogenesis is an important response to injury, it is generally distinguished from neural regeneration (Lichtman 1982). However, it is clear from the literature that what we call neural regeneration is not limited to the *exact* reproduction of original circuitry, and in some cases, repair that supports partial or complete recovery of function may be the best that can be hoped for.

1.1.4 Injury Tests the Limits of Neural Plasticity

Our understanding of the mechanisms called into play during neural regeneration has expanded along with our appreciation for the plasticity of the nervous systems of both invertebrates and vertebrates (Murphey 1986; Bulloch and Ridgway 1989; Weeks 1990). Normal neuronal function calls for adjustments in the strength of connections within a relatively stable system. Neural injury reduces the constraints that maintain stable synaptic connections and opens the way for more extensive remodeling.

In the nervous systems of many animals, the possibility of achieving a new balance in the circuitry among uninjured and regenerating elements is promoted by widespread sprouting and novel synapse formation. The regenerating neurites compete for targets in an environment that offers different choices from those that had been available during development. Each neuron is constrained by its differentiated state to a subset of the available targets, but its own survival depends upon formation of some appropriate connections.

1.2 Replacement of Neurons

1.2.1 Where Will New Neurons Come From?

Replacing part of a nervous system is a more demanding process than repairing body systems with simpler cellular differentiation and less interdependence. Major damage to any system involves cell loss; however, the nervous systems of most animals cannot compensate, as the liver can, by restoring functional capacity through enlargement of the remaining lobes through cell addition. Neurons are individualized by their derivation, connections, and the experiences reflected in those connections (Goodman and Shatz 1993).

In regeneration that involves replacement of neurons, missing cell types would presumably be replaced by neurons arising from neuroblasts. For most animals, whether neuroblasts exist and whether neural generation and differentiation can be promoted in adults is not known. In invertebrates such as rotifers and nematodes, the specification of cells during development is so rigid that it leaves no room for small pockets of undifferentiated cells that might serve as progenitors of replacement neural tissue; at the other extreme are the cindarians, in which the constant renewal of all body cell types assures neuron replacement. Most animals fall somewhere between these two extremes, with the theoretical possibility of neuron replacement but a low probability of its occurrence. More information about the processes that govern development may allow us to recreate conditions that favor generation of replacement neurons.

1.2.2 How Will New Neurons Find Their Way?

CNS repair that involves replacement of neurons requires not only appropriate mitotic activity but also a sequence of interactions, whether recapitulation of developmental steps or novel steps, which determine each cell's identity and interconnections. Neural development in both vertebrates and invertebrates typically involves guidance cells, either neural or nonneural, transient or permanent, which structure the environment and participate, often through electrical connections, in the differentiation of developing neurons (Jellies et al. 1987; Braun and Stent 1989; Klose and Bentley 1989; Palka et al. 1992; Goodman and Shatz 1993; Fig. 1.2). Later-differentiating neurons can often follow the pathway established by the "pioneer" neuron by selective fasciculation.

In contrast to the supportive environment and appropriately spaced cues that guide neurites during development, neurites projecting from replacement neurons must penetrate established territory to restore distant connections, potentially encountering scar tissue or stop-growth signals (Caroni and Schwab 1988; Bovolenta et al. 1991). We need to understand the mechanisms that govern this growth in the few animals known to regenerate portions of the brain.

1.2.3 Which Animals Can Replace Neurons?

Replacement of all or part of the nervous system has been documented in a few invertebrate phyla, including coelenterates, flatworms, annelids, gastropods and tunicates (Chap. 2). Some vertebrates exhibit cyclical replacement or facultative repair in sensory systems and in other animals, replacement of discrete por-

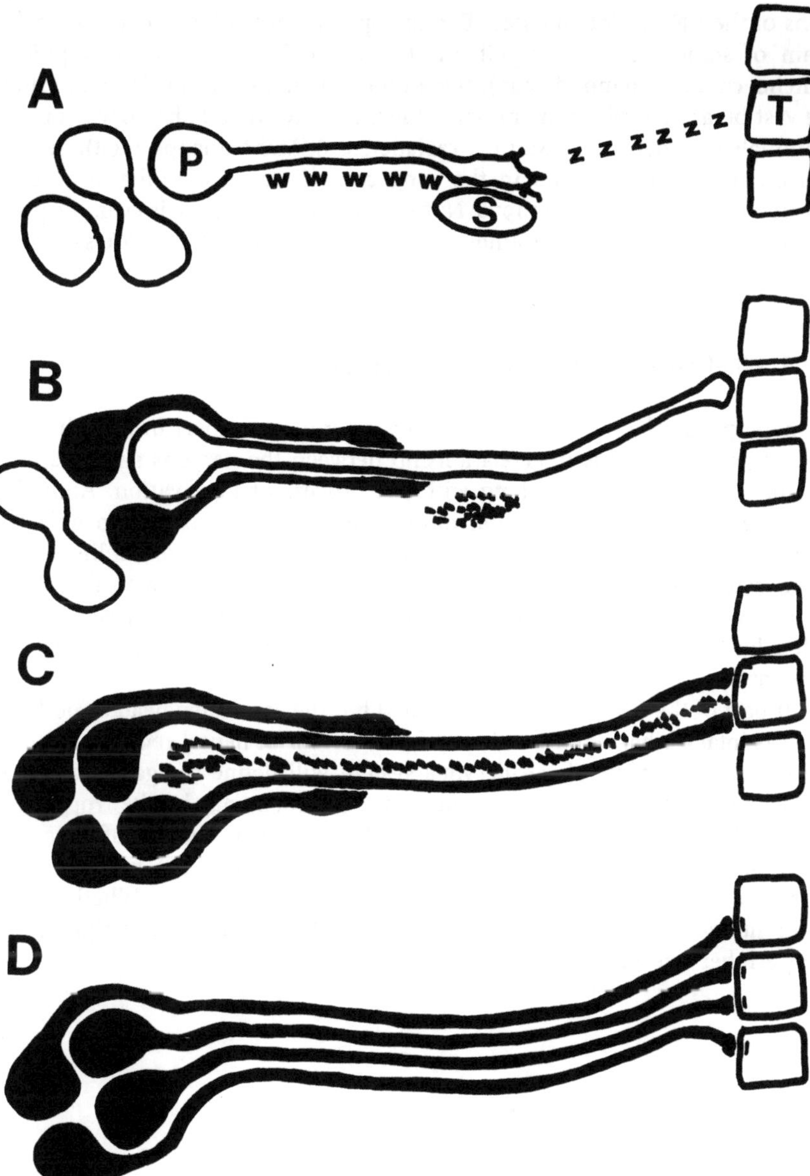

Fig. 1.2. A sequence of events seen in development of the nervous system of many animals. *A* Pioneer neuron (*P*) follows substrate cues (*wwwww*) to a signpost cell (*S*). In its interactions with the signpost cell, its affinities are altered, such that it is now attracted to a different substrate (*zzzzz*) that will lead it to a target (*T*). *B* Neurons follow the pioneer neuron by selective adhesion to surface molecules in its membrane. *C* The first neurons to arrive at the target form synapses. The pioneer neuron, like the signpost cell, may disappear. *D* Neurons differentiating later follow the membranes of similar cells and have second choices for target sites. Competition and neuron elimination (not illustrated) can occur at this stage

tions of the CNS is documented. For example, regeneration occurs in the forebrain of some fish and amphibians (Goss 1969), the retina of amphibians (Hitchcock and Raymond 1992), the auditory hair cells of birds (Travis 1992), the vestibular hair cells of mammals (Rubel et al. 1995) and olfactory neurons in mammals (Graziadei and Monti-Graziadei 1978; Mahanthappa 1994). Seasonal neuron replacement occurs in the song control nuclei of some bird species (Nottebohm and Nottebohm 1978; Nottebohm 1981) and there is even the potential for central neuron replacement in mammals (Reynolds and Weiss 1992).

1.2.4 What Factors Delineate Neurogenesis?

In many phyla, the nervous system plays an important part in inducing body part regeneration. This interrelationship between the nervous system and the rest of the body develops along with the process of innervation. Amphibian limb regeneration depends upon inputs from the nervous system only after innervation of the original limb has occurred, because an aneural limb can be regenerated without neural input (Singer 1959). An exception to this dependence on neural induction of regeneration is seen in insects, where clues provided by body gradients are largely responsible for organizing replacement elements.

If part of the CNS is injured or ablated but the associated head or body segment is intact, what signals will specify the details of neural regeneration? Perhaps remaining parts of the nervous system will define the gap and provide replacement elements. Another contributor to repair may be the trophic relationship between the nervous system and the periphery. The role of the periphery in directing differentiation of the nervous system is documented in the case of late-differentiating neurons in the leech (Baptista et al. 1990). Whether or not differentiation of replacement neurons might depend upon such signals from the periphery needs to be explored.

1.3 Regrowth of Neurites

1.3.1 Incidence of Neurite Repair

In contrast to the relatively small number of species known to exhibit central neuron replacement, many species repair central connections, and the repair of peripheral nerves is even more widespread. The nervous systems of most invertebrates, like those of vertebrates, are characterized by central collections of neuron cell bodies linked to the periphery by nerves. (Exceptions include the radially distributed nervous systems of ctenophores, cnidarians and adult echinoderms and the nematode pattern of target projections to centrally located neurons). Injuries that leave neuron cell bodies intact can often activate a repair program in those cells that leads to regrowth of the neurite and reestablishment of appropriate contacts.

Lower vertebrates regenerate portions of the CNS more successfully than mammals (Stuermer et al. 1992). However, the potential for CNS regeneration exists in birds and mammals and can be promoted in the mammalian CNS by introducing peripheral conditions. Manipulations that supply conducive pathways and avoid inhibitory effects of astrocytes allow the growth potential of CNS projections to be expressed (Aguayo et al. 1991). These studies illustrate that we must understand both the factors that promote regeneration and the factors that block it in order to gain control over the process.

1.3.2 Comparisons Between Development and Regeneration

During development, neurons are generated and arrive at their destinations in a particular sequence. The transient exposures to cell adhesion molecules expressed by cells such as glia, epithelia, and other neurites not only guide neurons along the proper pathway but may also trigger differentiation and lead to alterations important for target recognition. Competition for trophic factors derived from targets then determines process distribution and neuron survival.

Compared with developing neurons, regenerating neurons must locate and compete for targets under very different circumstances (Fig. 1.3). Regenerating neurites may be able to grow along the surviving or degenerating distal axon stumps and their glial investments in order to locate the target. Alternatively, pathway adhesion molecules and target-associated gradients may be present or reexpressed in regenerating animals (Stuermer et al. 1992; Palka 1993). However, the distances over which these factors must act are orders of magnitude larger in the adult than in the embryo. Neurons may be programmed to grow to a certain size or for a limited period, and if they do not locate a target operating under these constraints, regeneration may fail. A final complication can arise when replacement projections navigate to their targets only to find the synaptic space already occupied by other neurons.

1.4 A Cost-Benefit Analysis of Neural Regeneration

Neural injuries are likely to occur during the lives of most animals. Under what circumstances does regeneration pay off, i.e., under what circumstances will the commitment of energy to neural regeneration be likely to increase the reproductive success of that individual?

Neural function is so important that major disruption of the nervous system is usually lethal within a time frame that precludes repair. Injuries that

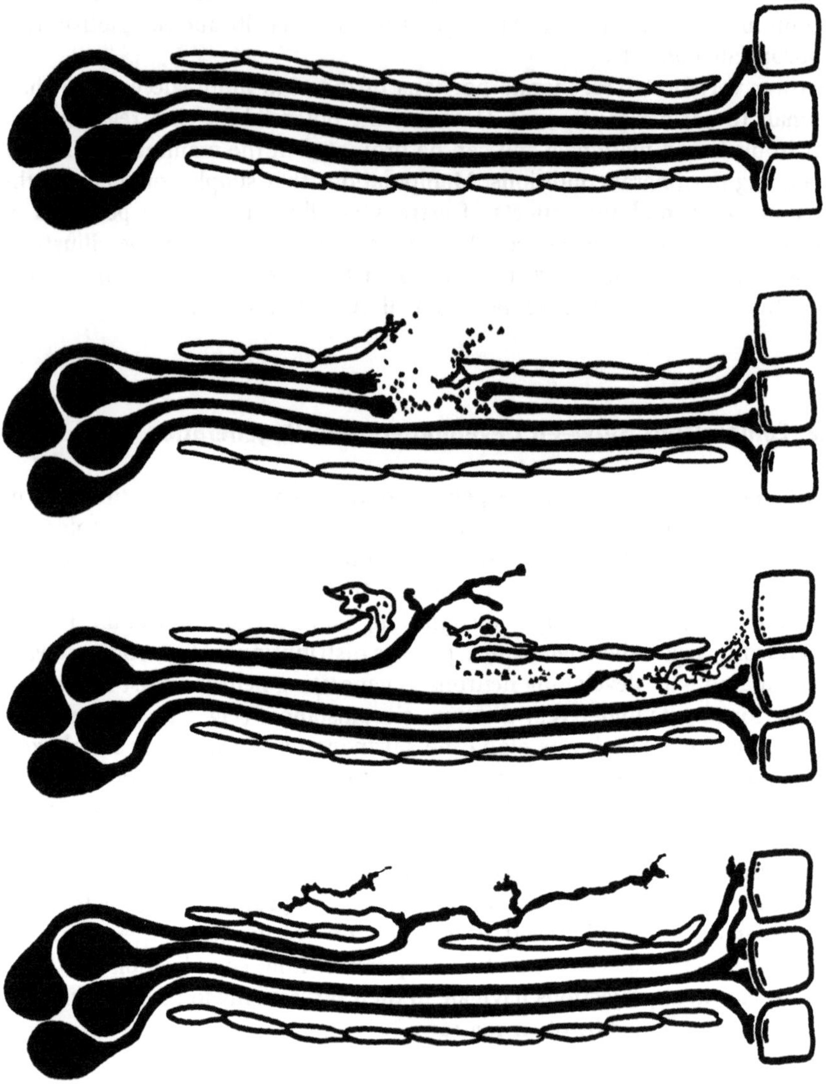

Fig. 1.3. Format is similar to that of Fig. 1.2. Prior to injury, the neurons are surrounded by a neural sheath. Upon injury, two of the neurons are axotomized. Phagocytic cells arrive on the scene and begin to clean up debris and promote repair. One axon goes astray, attracted to sheath material; the other injured cell locates the adjacent cell membranes and the degenerating distal stump. Guided by these cues and the nerve sheath, it arrives at the target area. However, one of the uninjured cells has sensed the vacated synaptic sites and has sprouted branches to occupy them. Perhaps the more successful of the regenerating axons will be able to form a substitute connection

compromise an animal's homeostatic functions, ability to escape from predators, or ability to compete for food would be lethal, so there would be no selection for the capacity to initiate repair following severe injuries. On the other hand, there appears to be a selective advantage to the regeneration of expendable body parts (limbs, body segments, tails, sensory appendages) whose injury or loss in encounters with predators is not lethal. The strategy of autotomy, or self-amputation, is seen in many animal phyla, and usually involves severance at a preferred site for fracture and optimization of regeneration from that site. For instance, if a crab's leg breaks at the autotomy fracture plane close to the body, blood loss is restricted and the subsequent regeneration reproduces the muscular, skeletal, circulatory and neural elements necessary to restore a complete limb; if breakage occurs either distal or proximal to the fracture plane, regeneration is imperfect (Bliss 1960; Goss 1969).

Replacement of a defined region of the body, such as a series of body segments in an earthworm, allows piecemeal reenactment of the developmental program. In this process, tissue interactions, polarity cues, and the timing of the appearance of neural elements coincides with that of many of their targets. Such regeneration must be distinguished from neural regeneration in the absence of body-part replacement, as stressed by Hulsebosch and Bittner (1980). In the latter situation, a mismatch of adult tissues and newly produced neurons introduces a time warp that may be difficult for the new neurons to bridge.

The younger an animal is at the time of the injury, the longer the commitment of resources can pay off in terms of competitive success. Immature animals generally regenerate more successfully than adults. This may correlate with the presence of growth-associated proteins in the immature neurons (Snider and Johnson 1989). Similarly, in developing nervous systems adjustments in cell fates can be made in response to neuron loss (White et al. 1991).

In contrast to young animals, reproductively active animals have the least time and resources to devote to neural or other body part repair. Some hormones that support reproduction are generalized growth factors which may also promote regeneration. However, the metabolic demands of reproduction are potentially in competition with the metabolic demands of regeneration.

The cold-blooded vertebrates exhibit better neural regeneration than birds and mammals. The regenerating neurites may have a better chance to regain their former synaptic sites in such animals, because the rate of reactive synaptogenesis is relatively slower than the reparative growth, compared with the situation in mammals. However, most of the interference with CNS regeneration in birds and mammals is due to stop-growth signals that arise from central glia (Chap. 4).

The life spans of some invertebrates are indeterminate, as they continually renew the body (Cnidaria), add individuals by budding (Platyhelminthes), or reproduce by fission (some Annelida, Cnidaria). In such animals, repair of the nervous system can be achieved with relative ease because cellular renewal is part of the ongoing growth pattern.

Addition of neurons throughout the life span potentially confers a developmental flexibility to the system involved. In many invertebrates and vertebrates, sensory cells are generated throughout life, and the retention of substrate cues used by added sensory neurons in locating CNS targets may facilitate incorporation of replacement sensory neurons following injury. In other animals, including mammals, totipotent cells or neuron precursors have been demonstrated. These confer the theoretical capacity to replace lost neurons, provided their mitogenic and trophic requirements can be met.

1.5 Rationale for Studying Invertebrate Regeneration

1.5.1 Invertebrates Offer the Simplest Systems

Invertebrate nervous systems have frequently been promoted as "model systems" on the basis of the relative simplicity of their nervous systems and presumed stereotyped behavior. This view has changed with the appreciation of the range of their adaptive responses, including learning (Murphey 1986). This revised view of invertebrate neural plasticity has gone hand-in-hand with the exploration of neural regeneration (Bulloch 1985; Bulloch and Ridgway 1989). Meanwhile, previously unsuspected capacities for repair have been recognized in the regeneration potential of neurons in the mammalian CNS. These findings have breathed new life into investigations of regeneration in vertebrate and mammalian nervous systems.

Significant gains in the understanding of neural regeneration will be made by studying this process in both the invertebrates and the vertebrates. Invertebrate animals and their nervous systems are fascinating in their own right, and deserve the appreciation that further study is bound to accord them. However, studying invertebrates has, in the past, led to a greater understanding of basic processes which were less accessible in vertebrates. Just as the squid giant axon provided the ideal preparation for early investigations of the action potential, the nervous systems of some invertebrates have been embraced for regeneration studies because their large neurons can be identified, isolated, and subjected to experimental manipulations.

Students of invertebrate neural regeneration are able to probe systems of identified elements with known connectivity, take them apart, and observe their reassembly under defined conditions. The complex phenomenon of regeneration can be approached most easily in a nervous system in which uniquely identifiable neurons allow the individual cell's responses to external cues to be analyzed at the level of receptors, internal messengers and induction phenomena. Given the common molecular language of neural differentiation, knowledge gained readily in simpler nervous systems may guide experiments aimed at improving the success of regeneration in the mammalian nervous system.

1.5.2 Invertebrates Offer a Great Diversity of Solutions

A second argument for studying invertebrates, which, after all, represent 95% of the animal kingdom, is the great diversity of their responses to neural damage. Some of these responses may suggest alternative strategies for repair that may be applicable to different parts of the complex mammalian nervous system. If, indeed, there is an ideal experimental subject for each question, invertebrate neural repair may reveal different strategies that may be applicable to the mammalian CNS, sensory receptor systems, peripheral nerves, neuromuscular function or autonomic neural plexuses.

Studies of neural regeneration in invertebrates have been focused on a relatively small number of species, chiefly those annelids, arthropods and gastropods whose nervous systems were already defined in behavioral, connectivity or developmental studies. It was logical to move from models of neural function to repair of known connectivity following neural injury. Unfortunately, most of these best-studied species exhibit a rather limited capacity for regeneration. These limitations have their parallels in the limitations seen in the mammalian condition, which may facilitate comparisons, but it may also have limited the scope of the questions addressed and the pursuit of dramatic and far-reaching solutions.

If our original goal had been to study those animals that regenerate most successfully, the choice of experimental material would undoubtedly have been different, as will be apparent from the survey in Chapter 2. Nevertheless, the invertebrates already studied offer some novel solutions. For instance, the quickest way to restore a severed nerve is for the individual proximal and distal segments of each axon to rejoin. Some invertebrates take this approach, which restores function rapidly (see Sect. 3.4), and experimental manipulations to promote such regeneration in vertebrates are being explored (Blundon et al. 1990; Krause and Bittner 1990). Without the example provided by invertebrates, it is unlikely that such a revolutionary approach would have been conceived. Additional model systems may be developed from among other invertebrates which could inspire research applicable to neural repair in human beings.

1.6 Conclusions

Repair of the nervous system entails a cost but potentially provides a benefit. Adults of a given species may or may not be able to activate appropriate regulatory pathways to promote repair. One of our goals is to understand these pathways in species in which they are operating successfully. Much can also be learned from the failures. With such information, we may eventually be able to circumvent the developmental constraints that limit neural regeneration in our own nervous system.

Some invertebrates are constantly renewing the body or parts of it, and others have only a brief period in which to make the most of the body they get. Where neural regeneration occurs, it is guided by some of the same signals utilized during development. Stress or injury release hormones and growth factors that promote growth and lower the constraints on neuron structure. In this permissive state, reorganization of connectivity among the existing neural elements may produce a functioning neural substrate to serve the organism's needs. That is the role of neural regeneration.

Chapter 2

A Survey of Neural Repair in Invertebrates

2.1 Introduction

The aim of this chapter is to provide a comparative overview of neural regeneration in invertebrates. Many phyla covered in an invertebrate zoology text are not mentioned in this overview because little is known, and I hope that the adventurous with access to such animals will be inspired to explore their regeneration potential.

An animal's responses to injuries encountered in nature are the real-life counterparts of responses to experimental lesions studied in vivo or in vitro. I have emphasized whole animal responses to neural injury in this chapter because the rest of this monograph emphasizes cellular and molecular approaches. Some descriptions of regeneration rest on chance observations or experiments aimed at understanding another phenomenon. A topic allied to neural regeneration, the role of the nervous system in inducing body part regeneration, was a focus of much of the older work.

2.2 Cnidaria: Neural Structure and Behavior

The body of cnidarians (coelenterates) is basically a tube consisting of an inner endoderm and an outer ectoderm. Polarity is organized relative to the oral pole (hypostome around mouth, surrounded by tentacles) and the aboral pole (the polyp's base or the uppermost part of the medusa's dome). In coelenterates, the epithelial cells seem to function in parallel with the neurons as a communication system. There is even evidence that epithelial conduction exhibits habituation (Grigoriev and Spencer 1994).

The radial symmetry of cnidarians is reflected in the nervous system, which consists of collections of neurons in nerve rings linked by nerve nets. The nervous systems of the simplest coelenterates, such as *Hydra*, are largely uncentralized networks, but the tentacle ganglia of hydromedusae are as complex as the bilaterally symmetrical brain ganglia of flatworms (Fig. 2.1; Mackie 1980).

Despite the absence of a central ganglionic mass, Cnidaria exhibit beautifully coordinated behaviors, such as escape swimming (Singla 1978; Spencer and Satterlie 1981), withdrawal activated by giant fibers (Mackie 1980; Roberts and Mackie 1980), tentacle autotomy (Bickell-Page and Mackie 1991), and behaviors involving complex neural integration (Haralson and Groff 1975).

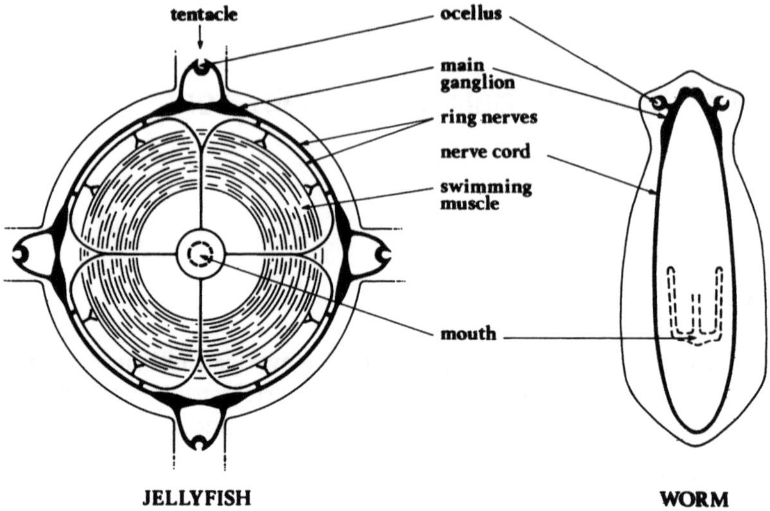

Fig. 2.1. The central nervous system of a jellyfish such as *Sarsia* compared with a planarian worm. (Mackie 1980)

Much of the interest in the synaptic systems and neurotransmitters of cnidarians stems from evidence that this is the oldest living group to possess a nervous system. Many classical and peptide transmitters found in other invertebrates and vertebrates are found in the cnidarians.

2.2.1 Examples of Regeneration

Some of the beautiful animals in this phyla have the capacity to regenerate an entire organism from a body fragment (Hay 1966). Corals and sea anemones (Anthozoa) and polyps (Hydrozoa) all exhibit excellent regeneration of body parts. Conduction of excitation can be rapidly reestablished across a cut or a graft in some members of the class Scyphozoa, but the mechanism of repair needs to be clarified.

The groundwork for future studies of neural regeneration has been laid by studies of cultured neurons and descriptions of differentiated neuron types and defined synaptic physiology in Cnidaria (Anderson and Schwab 1982; Anderson and Spencer 1989; McFarlane et al. 1989; Koizumi et al. 1989; Westfall et al. 1994).

Neurotransmitter immunohistochemistry has revealed specializations within the nerve net that characterize various body regions. The nerve net of the head of *Hydra* regenerates along with the new head; abnormal head regeneration can be studied by depleting the interstitial cells that give rise to neurons and replacing them with corresponding cells via a graft from another animal (Koizumi et al. 1990). These studies indicate that the epithelial cell environment directs the pattern of neuron differentiation. Factors such as head-activator of *Hydra* (Schaller et al. 1989, 1990; Fenger et al. 1994; Fuentes et al. 1994) or the proportion-altering factor of *Hydractinia* (Plickert 1989) mediate the communication between different regions of the body that determines neural differentiation (Sect. 2.2.2).

Neurons immunoreactive to antibodies against small peptide transmitters (RFamides) appear in the same pattern in tentacle development and regeneration in the hydromedusa *Polyorchis* (McFadden and Spencer 1992). The RFamide-immunoreactive neurons (neurons expressing peptides in the arginine-phenylalanine family) appear during the ectodermal thickening stage and increase in number to form a cup at the base of the tentacle bud. Treatment of hydromedusae with hydroxyurea removes both neurons and the precursor interstitial cells and is correlated with loss of the ability to regenerate tentacles. [However, see Marcum and Campbell (1978) for evidence that *Hydra* regenerates in the total absence of these interstitial cells.]

2.2.2 Regeneration in *Hydra*

Hydra studies have provided the most complete picture of coelenterate regeneration. *Hydra* can respond to lesions of the body with migration and transformation of cells derived from other body regions. A fragment of the whole body will transform into a smaller but perfect individual. This capacity for repair is undoubtedly related to the fact that the body is continually in flux, renewed by generation of new cells and cell sloughing at the extremities (Fujisawa 1989). In well-fed animals, 10–20% of the neurons are replaced each day (David and Gierer 1974). Neurons transform from one immunoreactive/functional type to another as they take up different positions in the body (Fig. 2.2; Guthrie 1985; Bode et al. 1988; Koizumi et al. 1990; Koizumi and Bode 1991; Bode 1992). Among the neuron types, distinct lineages have been identified (Yaross et al. 1986).

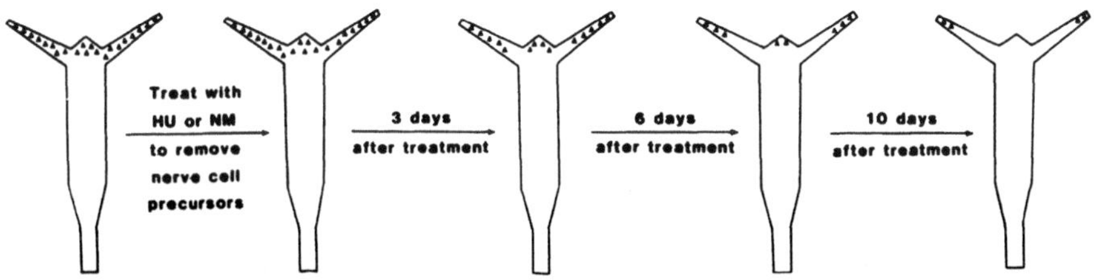

Fig. 2.2. Experiments illustrating the displacement of epidermal sensory neurons (*triangles*) along the tentacles with time in the absence of new nerve cell differentiation. (Bode et al. 1988)

Polarization in an oral-aboral axis and the proportions of the body are determined by diffusible, short-lived molecules that influence cell differentiation (Plickert 1989; Schaller et al. 1989). For instance, head activator acts at several points in the cell cycle: it serves as a mitogen for the epithelial stem cells, as a determinant to insure commitment of epithelial cells to the head pattern, and it allows final differentiation after decapitation removes the head inhibitor factor (Schaller et al. 1990). A peptide with an amino acid sequence identical to head activator is found in endocrine and nerve cells of mammals (Schaller et al. 1989; Fuentes et al. 1994), where it is likely to have a similar mitogenic effect.

Hydra reproduces asexually by budding, which involves both localized cell generation and migration (Webster and Hamilton 1972). This capacity is correlated with excellent body regeneration in several groups of invertebrates. Neurogenesis involves dividing intermediates (Bode et al. 1990) and has been the subject of chemical perturbation and stimulation studies (Greger and Berking 1991; Müller 1991).

Neuron precursors are induced by treatment of the whole animal with head activator (David and Hager 1994). If the intact animal is producing a bud or if body tissue is injured, the neuron precursors migrate to the site and differentiate as neurons. In contrast, in intact *Hydra* these unstable undifferentiated cells are shortlived, indicating that they must receive an as yet undescribed signal from sites of growth or repair in order to differentiate.

2.3 Ctenophora: Comb Jellies

The ctenophore nervous system, like the cnidarian nervous system, exhibits a radial organization. It consists of a subectodermal nerve net, elongated nerve-like plexuses bearing neuron somata located beneath the comb rows, a similar concentration around the mouth, and concentrations of neural somata at the aboral sensory pole (Hernandez-Nicaise 1973). The regenerative capacity of this group includes replacement of the apical sense organ (statolith organ) and regeneration of half the body or more. The beautiful fluorescent pattern associated with activation of the comb rows is controlled by the apical sense organ. Recovery of the coordinated luminescence of a comb row would make an attractive assay for neural lesion repair.

2.4 Platyhelminthes: The Simplest Bilateral Nervous System

The Platyhelminthes (flatworms) are flattened, bilaterally symmetrical, triploblastic, and unsegmented. The group includes turbellarian (freeliving) flatworms such as planaria, the trematodes and monogeneria (flukes), and the cestodes (tapeworms). The flatworm nervous system is bilaterally symmetrical. In the most primitive forms there is some cephalization and a network supplying the body; more advanced forms possess cerebral ganglia associated with anterior sense organs and nerve cords extending to the rest of the body (Fig. 2.3; Baguna and Ballister 1978; Koopowitz 1989).

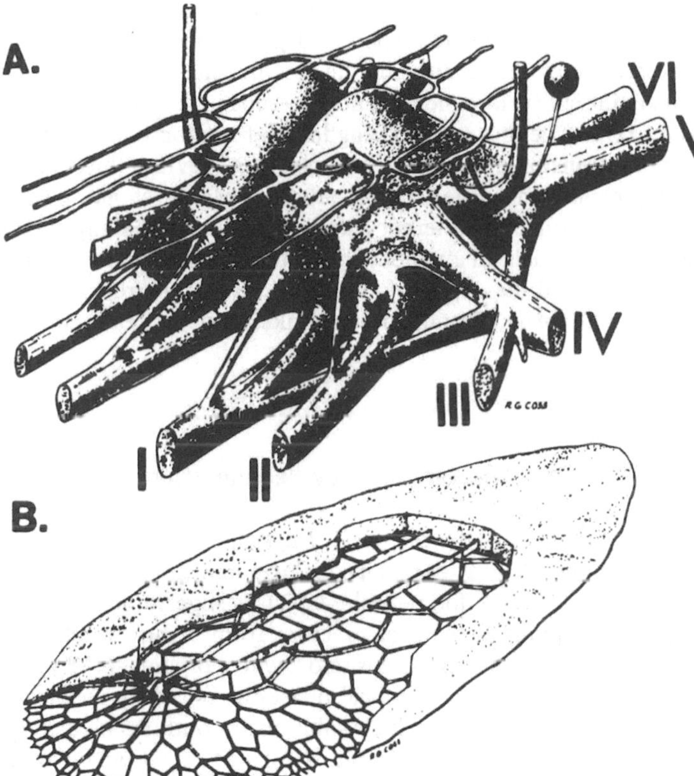

Fig. 2.3. **A** Serial reconstruction of the brain of *Notoplana acticola*. Drawing made from a model built up from 15 μm serial sections. The anterior points toward the *left-hand side* of the page. Clusters of globuli cells that lie outside the sheath, but that send their axons into the neuropile, can be seen at the anterior side; the finer dorsal submuscular plexus can be seen on the dorsal part of the brain. Nerves *I-VI* lead into the stouter ventral submuscular plexus. Note that a number of nerves exiting from the brain anastomose to form these nerves. The antlers exit from the lateral aspect and pass dorsally to innervate sensory tentacles. One eye is shown attached to the base of this nerve, but many eyes usually occur. This brain was approximately 600 μm in diameter. **B** The position of the brain and ventral submuscular plexus are shown in a cutaway of the animal. Note that the nerves are flattened-oval in cross section. The entire animal is approximately 25 mm long. (Keenan et al. 1981)

A large number of neuron types are revealed by neurotransmitter antibodies, suggesting that these molecules have been utilized from very early in animal evolution (Gustafsson and Reuter 1990; Gustafsson 1992; Gustafsson and Eriksson 1992). Glial cells are apparently absent from the nervous systems of some parasitic forms. A prominent feature of turbellarian nervous systems is the bilateral system of electrically coupled neurons (Elvin and Koopovitz 1994).

Larval amplification, a form of asexual reproduction, is seen in the parasitic forms. Reproduction by budding or fragmentation is characteristic only of turbellarians such as the planarians; their regeneration from spontaneously generated fragments is very similar to their regeneration following injury.

In tapeworms, continuous growth results in a chain of body segments that are temporally staged with respect to degree of maturation. The nervous system possesses slight thickenings of the longitudinal nerve cords suggestive of a brain, a fine subsurface plexus, and generous innervation of the scolex (holdfast structure) and genital region (Lyngdoh and Tandon 1992). Mapping of the classical neurotransmitter families in tapeworms provides the basis of comparative, developmental and regeneration studies (Maule et al. 1993). The presence within one animal of different stages of maturation along the animal's length make these animals an excellent choice for examining the activation of developmental cues and the effects of growth factors on neurons (Palladini et al. 1988; Baguna et al. 1989b; Gustaffson and Eriksson 1992; Reuter and Kuusisto 1992).

2.4.1 Planarians: Masters of Body Regeneration

Turbellarians such as the planarian *Dugesia* exhibit an extraordinary capacity for replacement of body parts, producing a new head with eyes and brain ganglia or (a special trick for neurobiologists?) supernumerary heads (Brondsted 1969). Not stopping at that, they recover from being cut into six or more fragments, each of which reorganizes into a whole organism (Martelly and Franquinet 1984). Some of the theoretical questions posed by regeneration and some experimental approaches to their solution are outlined by Baguna et al. (1988).

Following a lesion, a blastema is formed of totipotent rather than dedifferentiated cells (Baguna et al. 1988; Baguna et al. 1989a). This distinction can be made by irradiating the host and introducing defined cell types (Fig. 2.4). Regeneration involves both epimorphic (new cell proliferation) and morphallactic (existing cell migration) processes, with one or the other mechanism dominating in a given species (Baguna 1976; Palmberg 1986; Salo and Baguna 1989). The sequence of the appearance of neuron categories in regenerating heads is similar to that observed during the production of heads on new buds in asexual reproduction (Reuter and Palmberg 1989; Koizumi et al. 1990).

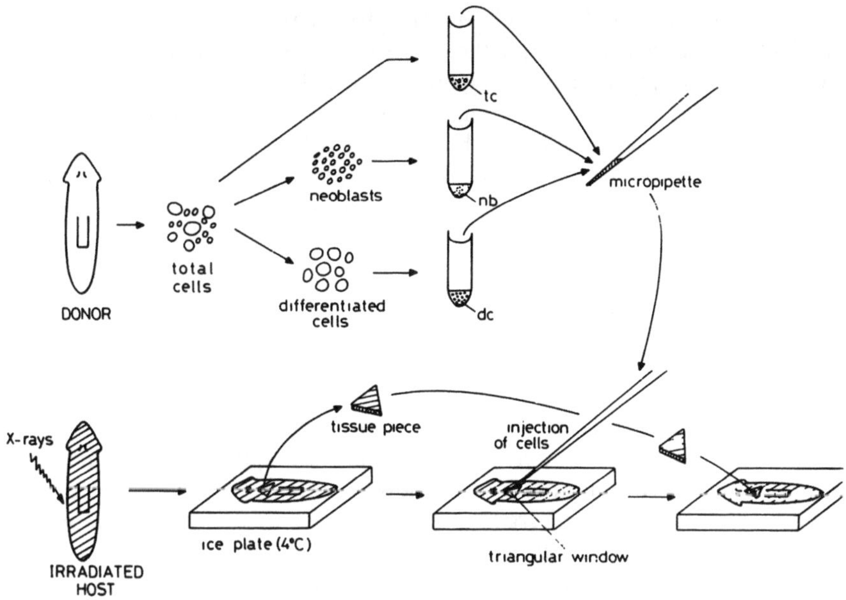

Fig. 2.4. The procedure employed to introduce total cells (*tc*) neoblasts (*nb*) or differentiated cells (*dc*) from nonirradiated donors into the parenchyma of an irradiated host (*hatched*). (Baguna et al. 1989a)

2.4.2 Role of the Nervous System in Body Repair

Several studies have implicated neurotransmitters in the control of planarian regeneration (Franquinet 1979; Martelly and Franquinet 1984). The concentrations of serotonin, dopamine and adrenalin in regenerating fragments increase, with serotonin and adrenalin elevated from 2–15 h and dopamine peaking at 15 h and again at 72 h. Addition of these neurohormones appears to drive synthesis of DNA and RNA, with Ca^{2+} and cyclic AMP suspected as second messengers (Martelly and Franquinet 1984). A number of other substances have been tested for mitogenic effects (Salo and Baguna 1986; Baguna et al. 1989b). The neuropeptide substance P, which is structurally related to bradykinins and *Hydra* head activator, exerts a generalized mitogenic effect on *Dugesia* tissues (Saló and Baguña 1986).

The nervous system provides polarity for the animal and is required for eye regeneration and gonad regeneration (Lender 1956; Ghirardelli 1965). Isolated postpharyngeal body segments lacking a portion of the nerve cord can regenerate a head, but posterior to the head there is no differentiation until a nerve cord is generated. In contrast, a complete worm can be generated from a body fragment, provided it has even a small piece of nerve cord, which apparently directs differentiation of other body tissues (Sperry et al. 1973).

In animals with a highly developed capacity for regrowth of any body part, it is of interest to determine how the process is normally suppressed. Experimentally, regeneration of a body region or internal organ can be inhibited by culturing lesioned animals in solutions containing extracts of the excised structure. For instance, regeneration of the pharynx is inhibited by a solution that contains extract of pharynx and bathing the animal in brain extract inhibits neural differentiation after brain excision (Ziller-Sengel 1965).

2.4.3 Repair of the Nervous System

Neural elements are replaced in the course of body part regeneration in the microturbellarian *Microstomum lineare*. The brain is quickly regenerated following its partial excision and the entire brain and eyes are regenerated after more extensive lesions (Palmberg 1986; Palmberg and Reuter 1990). Immunocytochemical staining reveals projections from the nerve cords to the site of regeneration. Neurotransmitter stimulation of regeneration is suggested by the immunocytochemical pattern during regeneration (Baguña et al. 1989b; Kimmel and Carlyon 1990).

Two homeobox-containing genes exhibit a concentration in the anterior region of the *Dugesia* body and are likely to be associated with neurons. They are expressed continuously but both show a slight rise in the middle of regeneration, when nerve cell differentiation is highest (Garcia-Fernàndez et al. 1991).

Polyclad flatworms have impressive powers of regeneration, although they are not as competent at replacing neurons as the planarians (Koopowitz and Holman 1988). Connections are regenerated following split-brain lesions in the triclad flatworm *Notoplana acticola* and lost behaviors are regained (Faisst et al. 1980). If one half of the brain is excised when the rest of the animal is uninjured, excised portions of the brain are not replaced by generation of new neurons. This is somewhat surprising, given the observation that neurons are added as the worms grow (Hauser and Koopowitz 1987). However, the nerves on the ablated side grow to the remaining part of the CNS and their connections with remaining nervous elements restore behaviors such as feeding (Koopowitz et al. 1975, 1976).

Blocks of body tissue containing the brain can be transplanted between animals (Davies et al. 1985). When the inserted tissue is rotated 180°, individual identified neurons are able to orient to the appropriate targets by alteration of their morphology (Fig. 2.5; Koopowitz and Holman 1988).

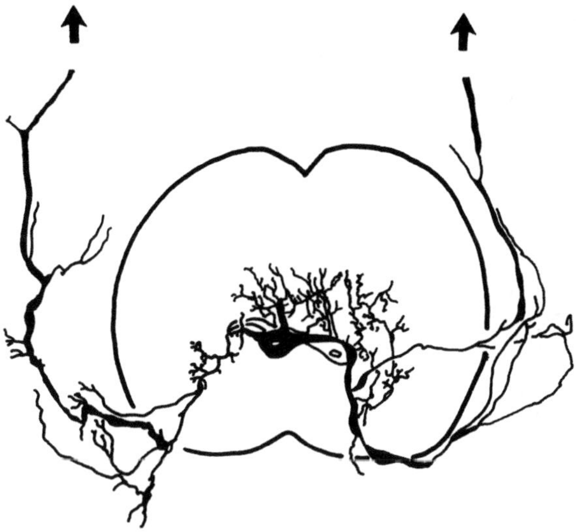

Fig. 2.5. *Camera lucida* drawing of the sBRA cells from the brain of *Notoplana acticola* after rotation of the brain through 180°, drawn from living preparation. Note that the axon from the unfilled cell *on the left* turns backward and grows by the brain to reach the target, while *on the right*, one of the dendrites is converted to an axon that grows to the target. The axons proceeding to the target have been cut short. Brain is 600 μm in diameter. (Koopowitz and Holman 1988)

2.5 Nemertea: Ribbon Worms

The nervous system of nemertine worms (Rhynchocoela) is more cephalized than that of flatworms. Eyes capable of light/shadow detection are present, but the animals depend primarily on chemosensory organs for tracking prey. The four lobes of the cerebral ganglia are linked by dorsal and ventral connectives which encircle the proboscis and associated structures. Ganglionated nerve cords (the number depending on the order) extend posteriorly, giving rise to mixed sensory and motor nerves.

Regeneration of body parts is highly developed in nemertine worms. These soft-bodied animals are frequently attacked by predators and the likelihood of body part regeneration is positively correlated with the likelihood of injury to that region. For instance, the eversible pharynx is frequently lost in predatory encounters and is readily regenerated.

Some nemertines can regenerate entire bodies from fragments, provided that the fragment contains a portion of the ventral nerve cord (Fig. 2.6; Hay 1966). As is true for planarian regeneration, specific parts of the body are not reformed if the regenerating system is grown in a solution containing an extract of that body part (Tucker 1959). Intriguing experiments on nemertines involving parabiosis and neural grafts are reviewed by Golding (1974a).

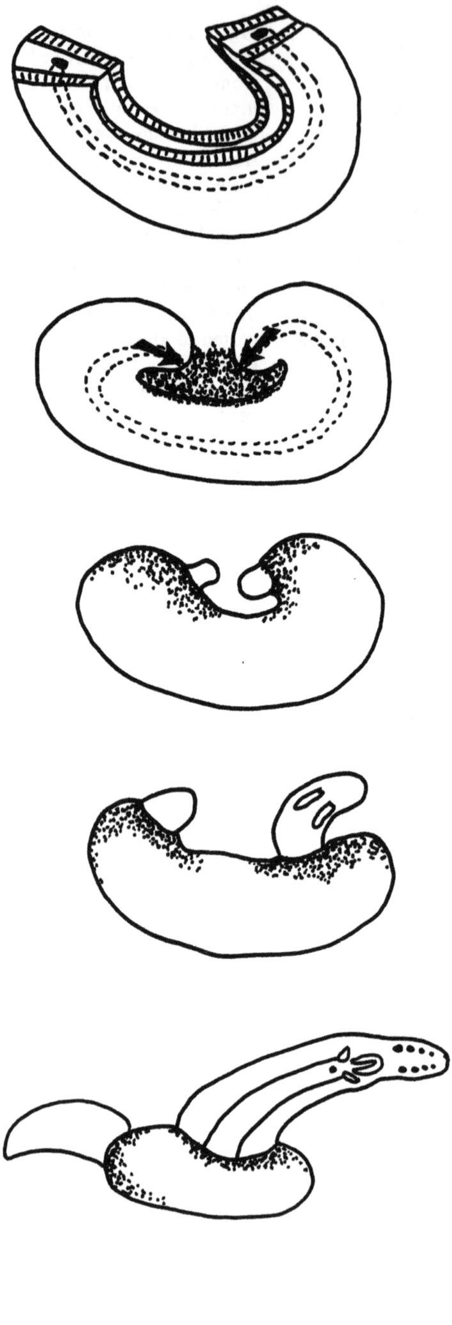

Fig. 2.6. Regeneration in a nemertine worm. Only those body-wall fragments that contain a segment of the ventral nerve cord initiate generation of a new worm from the fragment. (Redrawn after Hay 1966, originally from Berrill 1961)

2.6 Nematoda: Roundworms

Mitosis ceases in the bodies of roundworms when they reach maturity. Such animals lack stem cells that might serve as a source of new cells in regeneration. The nervous system of *Caenorhabditis elegans*, for instance, is defined down to the last cell division during embryogenesis and consists of 302 neurons (White et al. 1986).

Synapses between the neurons have been catalogued (Hall and Russell 1991) and the development of synaptic connectivity studied (Walthall and Chalfie 1988). Laser ablations provide clues to the function of particular cell groups (Chalfie et al. 1985). Elimination of individual cells has revealed a limited capacity for regulatory, as opposed to deterministic cell lineages, in nonneuronal cell lines (Kimble 1981).

Programmed cell death is an important part of neural development, and blocking this process reveals cells with the potential to differentiate as neurons. An example of neural equivalency is revealed in mutations which block programmed cell death (Avery and Horvitz 1987) and in some cases the 'undead' cells displace the function of normally surviving neurons (White et al. 1991).

Mutants reveal processes involved in cell migration and axon guidance and in the formation of connectivity (Hedgecock et al. 1990; White et al. 1992; Walthall et al. 1993). Signal molecules involved in pathway determination identified in *C. elegans* are closely related to those in insects and vertebrates (Chap. 4). Persistence of some of these signals might promote regeneration, but the extent to which the nervous system of adults can repair connections has not been explored.

2.7 Annelida: Introduction to the Phylum

The annelids are segmented worms. They comprise a diverse phylum with regeneration capabilities that range from good to excellent. This group includes the marine polychaetes, oligochaetes (earthworms), and the hirudinea (leeches).

The nervous system of annelids is characterized by a dorsal cerebral ganglion (actually, a pair of ganglia) that is more highly developed in those groups with specialized anterior sensory structures. The cerebral ganglion is linked by connectives to the subpharyngeal ganglion and the chain of ganglia along the ventral nerve cord. The serially homologous paired ganglia are present in each segment along the length of the body, and the number of ganglia can be very large in polychaetes and oligochaetes.

2.7.1 Polychaetes

Polychaetes are the most ancient of the three major groups of annelids, but among them we find the most highly developed nervous systems and sensory organs in the phylum. Some polychaetes and oligochaetes normally reproduce asexually by fragmenting, generating complete individuals from each fragment. Others can regenerate anterior and/or posterior body segments, or some portions of them, following injury.

Body part regeneration, where present, is dependent upon the nervous system. Neural control has been linked to the presence of a supraesophageal (cerebral) ganglion (Golding 1967a-d, 1974b, 1987; Olive 1974). In the polychaete *Nereis diversicolor*, Clark et al. (1962) used transplantations to document a 3-day interval between posterior segment removal and the release of a regeneration-promoting factor by the ganglia. The cerebral ganglia may not exert so strong an effect on posterior segment regeneration in all polychaetes (Pfannenstiel 1984).

The repair-promoting influence of the nervous system declines with age, as can be demonstrated by transplanting the cerebral ganglia from young animals into old animals and subsequently testing their capacity to induce anterior segment regeneration after decerebration (Golding 1987). The "aging" of the cerebral ganglia can be reversed by transplanting ganglia from older individuals into the bodies of young animals. This suggests that feedback from the aging body down-regulates trophic system(s) in the ganglia.

Many questions concerning the control of regeneration in polychaetes have been posed by the available observations. For instance, the relationship between a blood-borne factor released by the cerebral ganglia and the local control exerted by the ventral nerve cord is not clear; perhaps both are required. In the polychaete *Ophryotrocha puerilis*, the ventral nerve cord provides both inducing potential and positional information (Pfannenstiel 1984).

Disruption of the ventral nerve cord can lead to production of a supernumerary "tail" that has the same potential to grow and regenerate as the original body posterior segments (Pfannenstiel 1984). Determination of segment specificity by the ventral nerve cord is evident in experiments involving deflection of the ends of the severed ventral nerve cord to the body wall (Fig. 2.7; Goss 1969). This leads to the generation of a new, laterally projecting tail where the rear end of a segment of the cord contacts the body wall, and a new laterally projecting head where the front end of the severed cord contacts the body wall (Moment 1951; Kiortsis and Moraitou 1965). Other classical experiments on ganglion extirpation and grafts are described in the literature (Herlant-Meewis 1964; Golding 1967a-d; 1974; Goss 1969; Olive 1974).

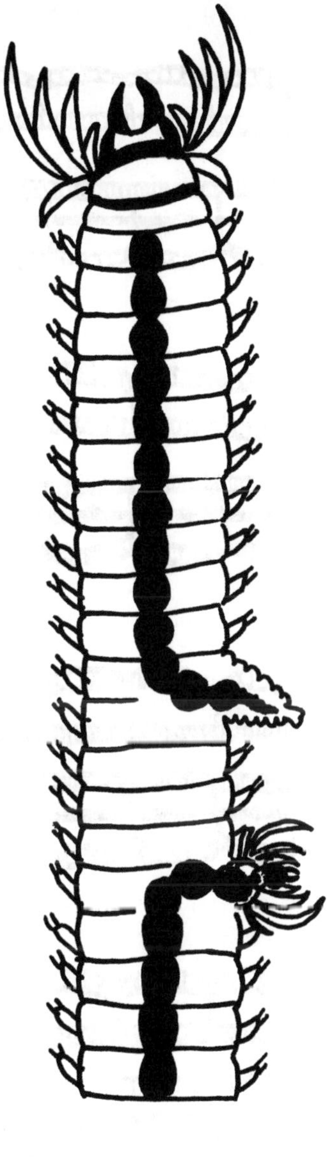

Fig. 2.7. Polychaete body segment induction by cut ends of the ventral nerve cord. (See Avel 1961 or Kiortsis and Moraitou 1965)

2.7.2 Oligochaetes

Oligochaete regenerative capacities run the gamut from complete body regeneration from fragments to much more limited body repair. The ability to regenerate body segments is reviewed by Drewes and Fourtner (1990). Regenerating neurites must grow into the dedifferentiated cell region to induce regeneration (Fig. 2.8). The relative contributions of undifferentiated cells, blood cells, and

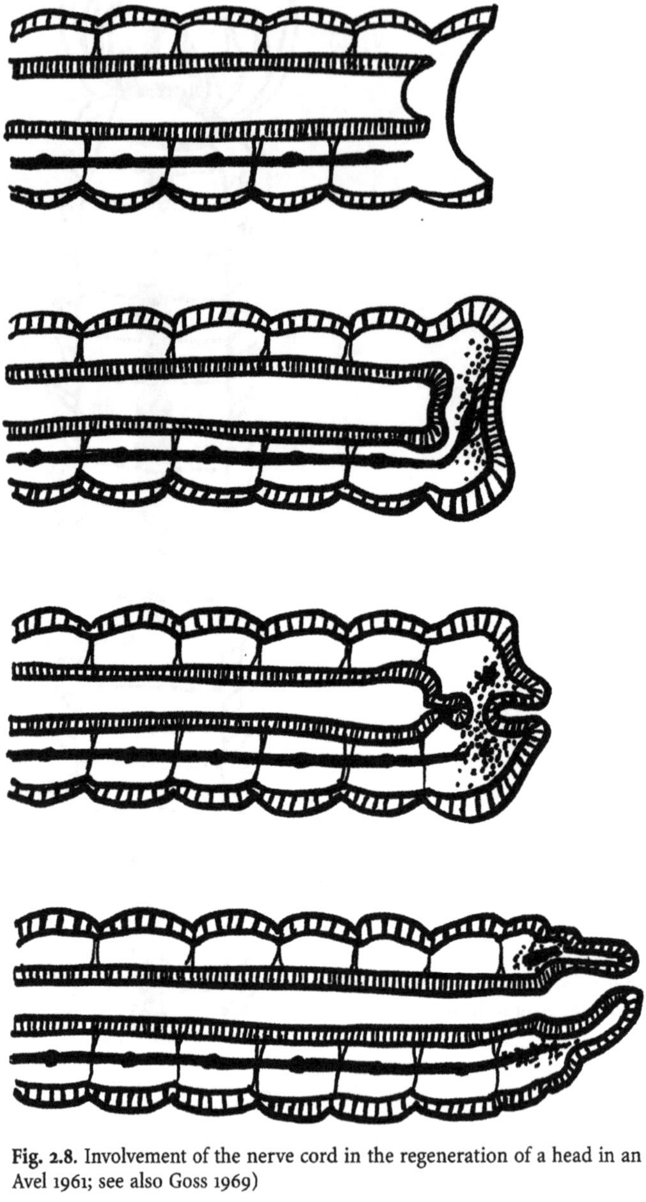

Fig. 2.8. Involvement of the nerve cord in the regeneration of a head in an earthworm. (After Avel 1961; see also Goss 1969)

dedifferentiated cells varies with the species (Clark et al. 1962; Herlant-Meewis 1964).

Ganglia transplanted into a host nervous system are incorporated into the body and mediate reflexes within 4–14 days. In the earthworm *Eisenia foetida*, the site of the origin of a 12-ganglion length of transplanted cord determines the reflex responses in the body segments receiving the ganglia. For example, transplantation of a length of anterior nerve cord into a posterior body location results in behaviors (direction of peristaltic shortening and setae protraction or retraction) appropriate for the site of ganglion origin but reversed for the body region where the ganglia reside (Vining and Drewes 1985a,b). This indicates that the ganglia retain circuitry specific to their site of origin but make appropriate connections with serially homologous sensory and motor elements in the body wall. Some plasticity is also demonstrated, as described below.

Responses of closely related species to particular lesions can be quite different. Some members regenerate neurons in the absence of body part regeneration whereas closely related species do not (Herlant-Meewis 1962; Hulsebosch and Bittner 1980). Regeneration of the giant fibers of earthworms and leeches can occur by fusion, recapitulating the process of formation of these giant fibers during development (Birse and Bittner 1976; Frank et al. 1975; Sect. 3.3). Behavioral recovery is correlated with anatomical evidence that the giant axons form connections across the lesion site; no inappropriate connections are detected.

After segments are lost, the giant fiber escape system is reorganized. The medial and lateral giant fibers are responsible for anterior or posterior body region withdrawal. They run the length of the body but are regionally specialized in size, conduction velocity, and receptive fields. Medial giant fibers receive tactile inputs in the anterior of the body and mediate anterior withdrawal; the lateral giant axons receive inputs from receptors in the posterior of the body and mediate posterior withdrawal. A body fragment derived from the posterior region of the freshwater oligochaete *Lumbriculus variegatus* will regenerate few anterior segments and many posterior segments, producing a new body while effectively shifting a posterior section of the old body into an anterior position. Noninvasive recording from the giant fibers allows monitoring of giant fibers during recovery (Fig. 2.9; Zoran and Drewes 1987; Drewes et al. 1988; Drewes and Fourtner 1990). Within a few days, functional properties of the giant fibers are reorganized to correspond to their newly assigned positions, a process that begins even before addition of segments (Drewes and Fourtner 1990).

Another example of morphallaxis is seen in the common earthworm *Lumbricus terrestris*: the dorsoventral flattening behavior characteristic of posterior segments is shifted to more anterior segments following posterior segment removal. If the ventral nerve cord is interrupted by removal of a length of several ganglia from an otherwise intact worm, there is a tendency for the flattening behavior to become established in the segments anterior to the lesion and then to disappear in those segments when regeneration of the cord reestablishes connections with the more posterior segments (Gras 1984).

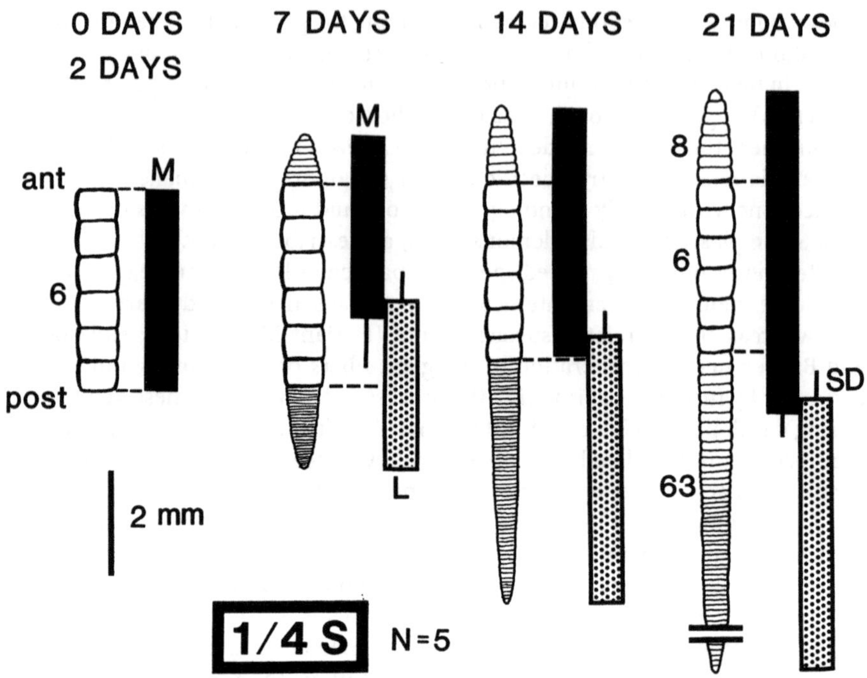

Fig. 2.9. Sensory field reorganization in short anterior fragments (group *1/4S*). Initially, all segments are represented by the medial giant fiber field, but at 7 days the lateral giant fiber field transiently emerges in two or three posterior segments of the fragment. Later, as the new tail section elongates, the medial giant fiber field expands to include all original fragment segments and several segments of the new tail section. (Drewes and Fourtner 1990)

2.7.3 Hirudinea

The leeches are the best-studied annelids from the standpoint of neurogenesis, neural connectivity and regeneration (Muller et al. 1981; Nicholls 1986; Loer et al. 1987; Stuart et al. 1987; Torrence et al. 1989; Thorey and Zipser 1991). Their phylogeny and behavior are reviewed by Sawyer (1981) and their development by Stent et al. (1992). The most commonly studied leech is the medicinal leech, *Hirudo medicinalis*. However, the nervous systems of others, including *Heamopsis marmorata* and *Helobdella triserialis* have contributed to the accumulating knowledge of this group (Johansen et al. 1985).

Leeches have a more limited body regeneration and do not exhibit the asexual reproduction common in the oligochaetes. They do not replace neuron somata, but their neural regeneration includes repair (fusion) of giant axons (Frank et al. 1975), regrowth of severed axons, reinnervation of the periphery, and reestablishment of central connections (Nicholls and Van Essen 1974; Mason and Muller 1983; Muller 1988).

The intensive study of leech neural development has allowed fruitful comparisons between the processes of development and regeneration (Macagno et al. 1989; Cole et al. 1989; Thorey and Zipser 1991; Muller and Gu 1991; Masuda-Nakagawa and Nicholls 1991). During development, a variety of interactions between neurons and nonneural cells modulate neuron fate (French and Kristan 1994). The target is instrumental in the final differentiation of some leech neurons (Gao and Macagno 1988; Merz and Drapeau 1994) whereas regenerated neurons are often insensitive to the presence of the target (Muller et al. 1987). Regeneration of the electrical synapse between S cells is independent of cues from the target (Muller and Carbonetto 1979). Growth of a regenerating S axon ceases at the site where the target cell should be, whether or not the target axon is present (Muller and Scott 1979). A different pattern for these cells is seen during development, when S neurons initially overshoot their site of synapse formation (McGlade-McCulloh and Muller 1989).

Regeneration following connective lesions involves widespread changes in the strength of connections far from the lesion site (Jansen and Nicholls 1972). Recovery of swimming behavior occurs in the absence of neural regeneration after ventral nerve cord lesions (Baylor and Nicholls 1971), a reminder that behavior may not reflect circuit repair.

2.8 Arthropoda: Neural Organization and Repair

This discussion will be limited to crustaceans and insects, the best-studied groups. The arthropod nervous system is similar to the annelid plan; head (cerebral) ganglia are linked by connectives running from the cerebral ganglia around the esophagus to a chain of ventral segmental ganglia. Govind (1992) illustrates the microscopic anatomy of the crustacean nervous system; larval neurogenesis in arthropods is illustrated by the example of the spider crab (Harzsch and Dawirs 1994). Anderson et al. (1980) compare arthropod development and regeneration with that of other invertebrates.

The arthropod body is regionally specialized (e.g., cephalothorax and abdomen or head, thorax, abdomen), but the ganglia located at different levels are sufficiently homologous in their architecture and chemical cues that regenerating axons can identify appropriate pathways when introduced into ectopic locations (Chap. 5). The motoneurons of ectopically implanted insect ganglia locate and innervate appropriate leg muscles (Bodenstein 1957) and if legs rather than ganglia are transplanted, the neurons of the host segment will innervate them (Young 1972).

As a group, the arthropods exhibit a more limited range of regenerative capabilities than the annelids. There is no evidence that CNS neurons are replaced after ganglion ablation (Bittner et al. 1974; Seshan and Bittner 1987). The capacity for regeneration of nongiant axons in the ventral nerve cord declines with age, and the limited regenerative response of giant axons is inadequate to support axonal fusion (Seshan and Bittner 1987).

2.8.1 Reinnervation of Muscle

The parsimony of the motor innervation in insects and crustaceans makes them ideal experimental subjects for studying the specificity of innervation. Arthropod muscles may be innervated by as few as one to three neurons, and these neurons are identifiable on the basis of firing pattern, location in the CNS, and axonal branching pattern (Young 1973).

Insect and crustacean motoneurons respond quite differently to nerve section or crush (Fig. 2.10). In insects, the portion of the severed axon not connected with the soma degenerates, whether it is a sensory neuron (soma typi-

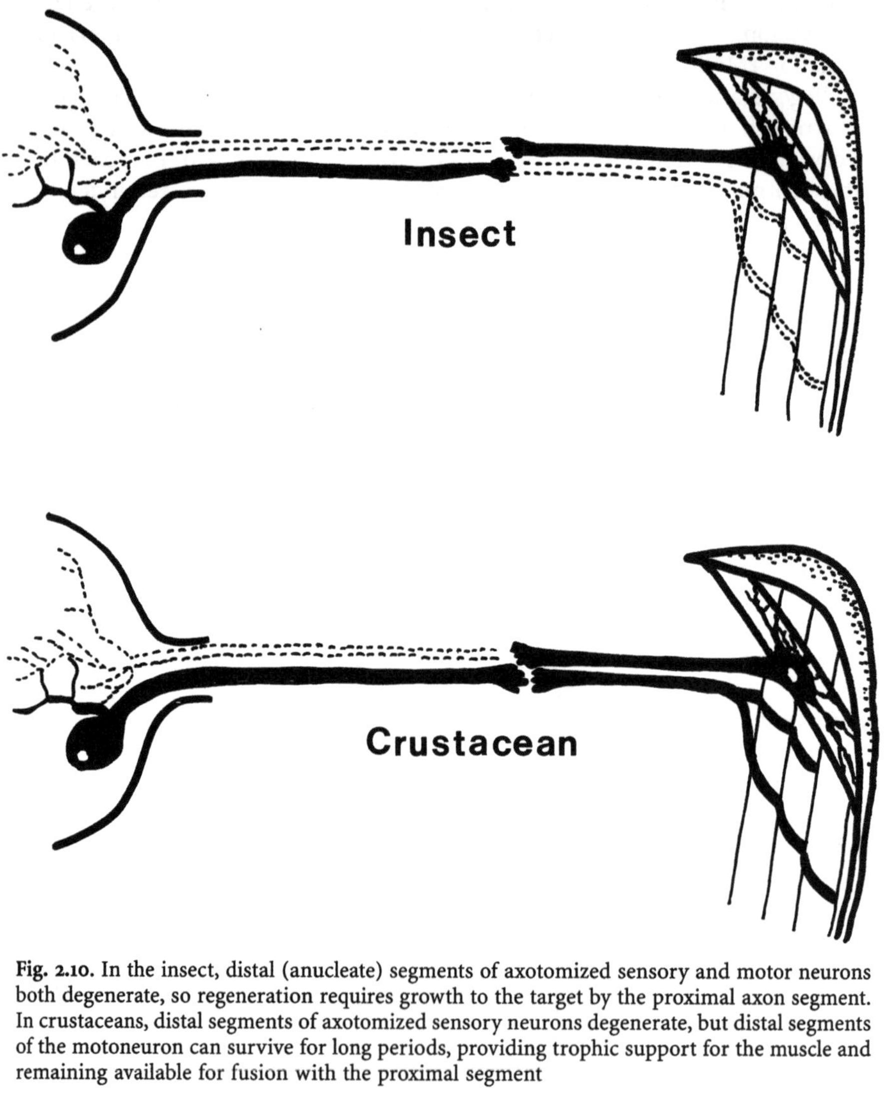

Fig. 2.10. In the insect, distal (anucleate) segments of axotomized sensory and motor neurons both degenerate, so regeneration requires growth to the target by the proximal axon segment. In crustaceans, distal segments of axotomized sensory neurons degenerate, but distal segments of the motoneuron can survive for long periods, providing trophic support for the muscle and remaining available for fusion with the proximal segment

cally in the periphery) or a motoneuron (soma in the CNS) (Guthrie 1967; Jacklet and Cohen 1967). In crustaceans, the distal portion of motoneuronal axons survive for many months after disconnection from the soma, whereas the anucleate sensory axon stumps degenerate (Bittner and Johnson 1974; Govind et al. 1992).

Due to the survival of the distal stump, regenerating crustacean motoneurons can sometimes form functional contacts (fusion or electrical continuity) between the growing end of the proximal (cell-attached) portion of the axon and the surviving distal stump of the same axon (Hoy et al. 1967; Chap. 3). This allows rapid restoration of communication with the periphery, and function is reestablished simultaneously for muscles in different limb segments innervated by the same excitor motoneuron (Bouton and Bittner 1981).

When fusion occurs, the muscles are actually reinnervated progressively (Fig. 2.11). The first stage is the engulfment of the distal end by many satellite axons. These contacts presumably provide sufficient electrical communication between the membranes to activate the stump by current flow (Bouton and Bittner 1981). After establishing effective contact with the distal stump, the satellite axons continue to grow and eventually establish chemical synaptic contacts with the muscles, at which time the distal stump degenerates. If the proximal portion of a motoneuron fails to contact the distal stump, it is capable of rein-

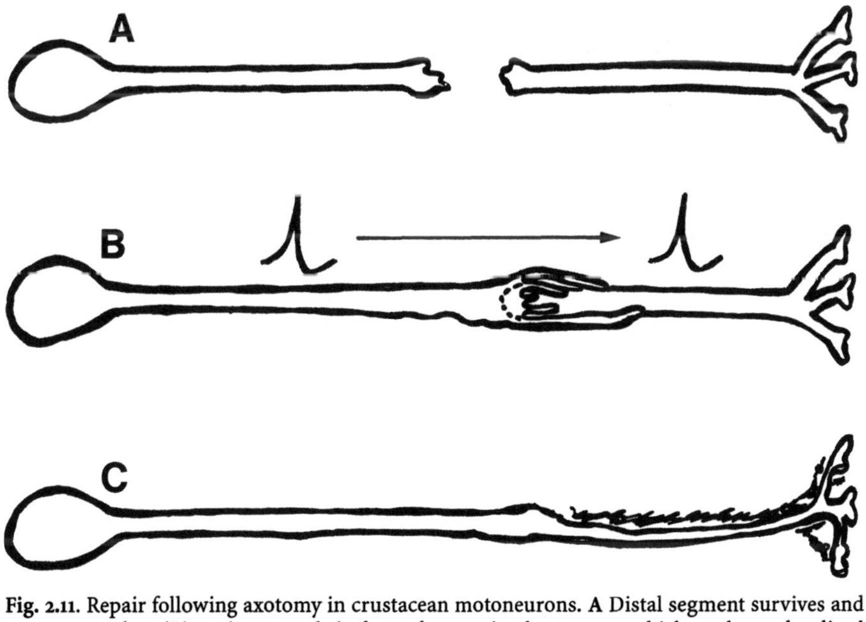

Fig. 2.11. Repair following axotomy in crustacean motoneurons. **A** Distal segment survives and may sprout, but (**B**) major growth is from the proximal segment, which encloses the distal stump in a basket of contacts, reestablishing electrical continuity. **C** One of the neurites continues to grow along the distal stump to the target and forms synapses on it; the surviving distal stump then degenerates

nervating the muscle by growing out and forming chemical synapses with the target; this mechanism of regeneration is ordinarily seen in some abdominal motoneurons (Bittner 1977).

In the superficial flexor system of the crayfish, the basis of specificity in muscle reinnervation has been investigated in a situation that precluded axonal reconnection (Ely and Velez 1982; Hunt and Velez 1982; Goransson et al. 1988). Regrowth to the periphery is followed by hyperinnervation and subsequent pruning of synapses, with individual differences between the identified neurons in their tendency to refine their connectivity.

2.8.2 Refinement of Connectivity

In the regeneration of muscle in some insects, there is exuberant development of motor projections followed by recognition of correct targets (e.g., Myers et al. 1990; Denburg and Caldwell 1992; Keshishian 1993; Chap. 5). Similarly, in the CNS of larval prawns, overproduction of connections is followed by loss of connections to form the mature pattern. Ablation of selected interneurons or motoneurons by irradiation in the larvae rescues synaptic connections among remaining neurons that are normally lost (Friedlander and Levinthal 1982).

2.8.3 Molting and Regeneration

The periodic punctuations of the life history of arthropods by molts, including the metamorphic transitions, involve changes in the nervous system that are coordinated by hormones (Weeks and Truman 1986). Hormones also control neurogenesis from undifferentiated cells in the mushroom bodies of the adult house cricket, *Acheta domesticus* (Cayre et al. 1994). In metamorphosis, some neurons die and others are reorganized to be integrated with neurons generated from embryonic neuroblasts (Weeks and Ernst-Utzschneider 1989; Truman 1990; Ito and Hotta 1992; Weeks et al. 1992). The situation of the surviving neurons has been likened to the challenge faced by surviving neurons regenerating in the aftermath of neuron loss due to injury (Briedbach 1988). During metamorphosis, an antigen present in the *Drosophila* nervous system during development is reexpressed in the extracellular matrix, presumably aiding axonogenesis and synaptogenesis (Go and Hotta 1992).

Regeneration of appendages is coordinated with the molt cycle (Needham 1965; Knobloch and Steel 1988) although regenerative growth of neurons continues during the intermolt. Autotomy, the intrinsically controlled severence of a limb at a fracture plane (Stasek 1967; Emson and Wilkie 1980), is widespread among arthropods (Bliss 1960) and can reset the timing of the molt by acting on the underlying hormonal systems (Skinner and Graham 1972; Hopkins et al. 1979; Skinner 1985; Knobloch and Steel 1988). The ability to regenerate a limb following autotomy is lost in adult insects but can be restored by grafting a prothoracic gland (ecdysone-releasing tissue) from an immature individual.

Autotomy in crabs involves activation of a high-threshold response by motoneurons, that severs the limb and allows escape from a predator (McVean 1975; Moffett 1975). Autotomy is also exhibited by arthropods that do not regenerate the lost part, but merely benefit by escaping with their lives. Degenerative changes are seen in the uninjured muscles that lie within the thorax, and changes occur in the central neuropil following autotomy which are likely to reflect alterations in neural input (Moffett 1987; Arbas and Weidner 1991; Schmiege et al. 1992; McTeague and Govind 1993). Comparisons between the motoneuron distribution of regenerated limbs and control limbs indicate that limb replacements, while functional, do not perfectly reproduce the original structures (Denburg and Whitington 1978). In addition to subtle differences in motor innervation, sensory neuron numbers are never completely restored in a regenerated limb (McTeague and Govind 1993).

If the basal limb ring of an autotomized crayfish limb is immediately replaced by a transplanted ring from another leg segment (cheliped or walking leg), factors that influence the nature of the regenerate can be observed (Mittenthal 1980). The cheliped site promotes more rapid growth in all grafts, which is correlated with the presence of a larger nerve in that site. Antagonism between a cheliped and cheliped tissue regenerating in an adjacent site is strong, and chelipeds also reduce the growth of adjacent walking leg regenerates. These experiments also produce some mosaic limbs and homeotic (biramous) limbs.

2.8.4 Regeneration of Sensory Cells

Like the annelids, arthropods add sensory neurons at each molt during postembryonic development but maintain a constant number of motoneurons and interneurons (Letourneau 1976; Hildebrand 1985). The fact that sensory pathways are repeatedly remolded during development to accept the intercalated elements is reflected in the fact that there is no critical period for the formation of these synaptic contacts (Edwards and Palka 1971; Chap. 5). Sensory neurons may be regenerated (Govind et al. 1988), following the pathways provided by existing sensory cells; however, the projections may also enter the CNS by different routes and bypass implanted ganglia to locate the host's own posterior ganglion (Edwards and Palka 1971; McLean and Edwards 1976).

Frequently, regeneration of an appendage results in homoeotic regenerates, i.e., replacement of one type of appendage with another, such as a leg in the place of the antenna (Bateson 1894). These naturally occurring instances of misplaced sensory cells, together with homoeotic mutants and supernumerary mutants, provide us with opportunities to learn what cues within the ganglia can be read by ectopically positioned sensory cells (Edwards 1988; Edwards et al. 1989). General pathway cues are read correctly by the ingrowing sensory cells, but specific targets may not be recognized when approached from ectopic locations.

2.9 Mollusca: Unsegmented Animals with a Range of Cephalization

The nervous systems of molluscs differ so greatly over the phylum that in their monumental work on the invertebrates, Bulloch and Horridge (1965) departed from their accustomed format to discuss the neuroanatomy of each class separately. The brains of cephalopods rival those of lower vertebrates in complexity; at the other extreme, the distributed nervous systems of amphineurans and monoplacophorans appear much more primitive. A peripheral nerve net and a degree of local autonomy is a characteristic of all classes.

The regenerative capacity of bivalves includes repair of structures that might be damaged by shell closure (the siphon or foot) (Needham 1952). The central nervous system of bivalves is well protected from injury in the natural environment as well as in regeneration experiments of scientists. In contrast, the relatively unprotected body of most cephalopods leaves them subject to extensive injury, despite a protected braincase. Cephalopods rapidly regenerate arms, including their rich endowment of peripheral neural elements (Lange 1920; May 1933). They repair peripheral nerves and restore neural control of chromatophores in the skin (Young 1932). Much remains to be learned from this group.

2.9.1 Gastropods: The Snails and Slugs

The nervous system of the Gastropoda falls between the extremes seen in the other classes. It is organized around a discrete number of central ganglia which are joined by commissures and connectives and which exhibit varying degrees of cephalization and fusion. Paired ganglia include the cerebral, buccal, pedal, pleural, and parietal ganglia. Additionally, the supraesophageal, subesophageal and visceral may be found alone or fused with other ganglia, resulting in ganglia that may bear the same name in different species but which are only partly homologous (Bullock and Horridge 1965).

Gastropod nervous systems have figured prominently in the recent literature on neural regeneration (Moffett 1995), so it is interesting to note that gastropods are not included in older surveys (e.g., Kiortsis and Trampusch 1965; Hay 1966; Goss, 1969; Eichler et al. 1972). A wide range of regeneration capabilities has been documented in the relatively few species studied. Incidental observations exist for many more, describing the ability to regenerate frequently injured body parts such as tentacles and the penis. Gastropods can also repair peripheral and central axonal projections and produce compensatory central circuits following ganglion lesions.

The vast majority of gastropods are marine, and the prosobranchs, the largest and most ancient group, have been the least studied. Evidence from a few prosobranchs suggests a well developed capacity for neural regeneration.

For instance, studies of the slipper shell, *Crepidula fornicata,* focused on neurosecretory phenomena but incidentally revealed a high tolerance for ganglion ablations and transplantations (Lubet 1971; Le Gall 1983; Le Gall et al. 1983). The nervous systems of two other major groups of gastropods, the pulmonates and opisthobranchs, appear simpler than the prosobranch nervous system. They possess fewer neurons and include some large neurons identifiable from animal to animal.

Studies of development in gastropods include the use of ultrastructural analysis (Page 1992a, b; 1993; Carroll and Kempf 1994), immunocytochemical probes (Kempf et al. 1992), and cell lineage studies (Jacob 1984; Cash and Carew 1989; Hickmott and Carew 1991) and the culturing of embryonic snail neurons will allow comparisons with the regenerative responses of adult neurons (Goldberg et al. 1988; Goldberg et al. 1991; Goldberg et al. 1992).

In some gastropods, neurons are added to the CNS during postembryonic development or throughout life (May et al. 1987; Roubos et al. 1988; Croll and Chiasson 1989; Cash and Carew 1989; Hickmott and Carew 1991). The gastropod pattern for neural development during body growth is somewhat similar to that for arthropods and annelids, in that some neurons appear early in development as identifiable cells (chiefly motoneurons and interneurons) whereas neurosecretory and sensory cells are added as an accompaniment of growth and/or age (Fig. 2.12).

Olfactory neurons of the snail tentacle are continually replaced, as they are in the mammalian olfactory system (Chase 1986; Chase and Rieling 1986). Some increase the number of interneurons also occurs in *Aplysia californica* during postembryonic development, contributing to increased neural plasticity (Rankin and Carew 1987). However, the identifiable giant neurons of opisthobranchs and pulmonates typically meet the needs of an increasing target size by increasing in size, selectively replicating portions of their chromosomes to support the necessary synthesis (endoreplication; Chase and Tolloczko 1987; Gillette 1991). Increases in size and process complexity are the sole mechanisms for matching the CNS with the growth of the periphery for both the mesocerebral and metacerebral neurons of the snail *Helix* (LaBerge and Chase 1992). Other categories of neurons increase somewhat in size, but primarily keep pace with the growth of the body through an increase in number.

2.9.2 Responses to Naturally Occurring Injuries

The shell, where present, protects the central nervous system; however, gastropod tentacles are frequently injured and commonly regenerate. When a stylommatophoran pulmonate regenerates its tentacle, this process can include regeneration of an eye and/or a tentacle ganglion (Flores and de Iraldi 1973; Chase and Kamil 1983; Flores et al. 1992). Different categories of neurons distinguished in normal tentacle ganglia are also found in the regenerated ganglion (Fig. 2.13). Moreover, olfactory function is restored in regenerated animals (Chase and Kamil 1983).

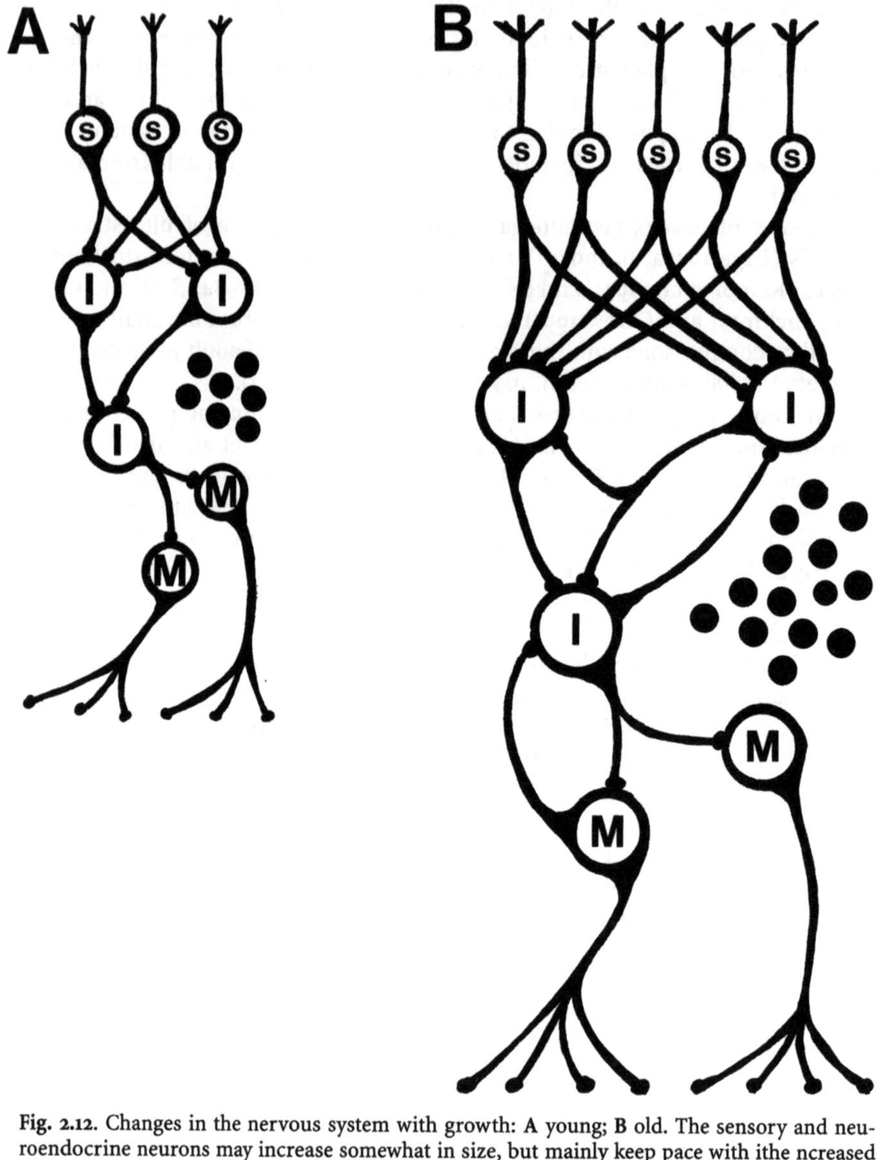

Fig. 2.12. Changes in the nervous system with growth: **A** young; **B** old. The sensory and neuroendocrine neurons may increase somewhat in size, but mainly keep pace with ithe ncreased size of the animal or emerging demands, such as reproduction, by an increase in number. Within the CNS, the motoneurons and interneurons may be constant in number, but increase in size and in the complexity of their processes

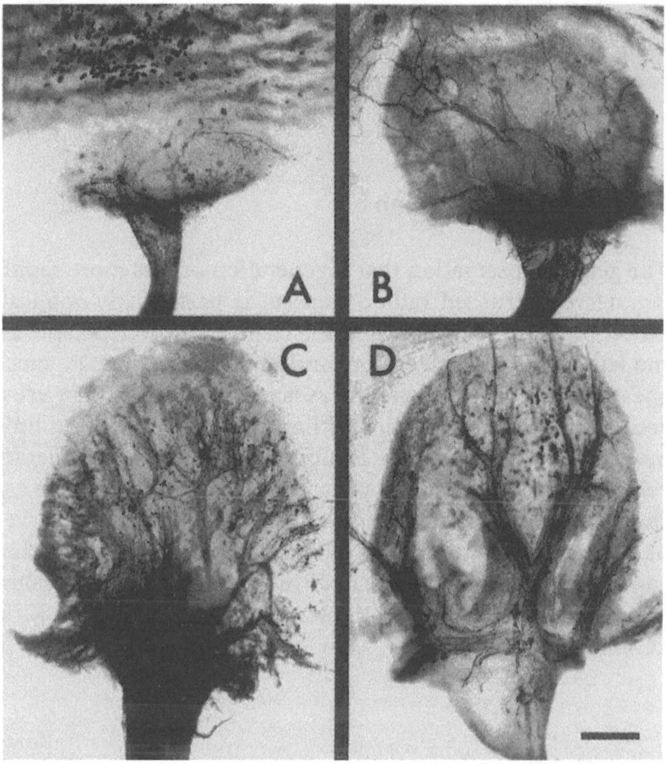

Fig. 2.13. Whole mounts of the tentacular olfactory structures at several stages of regeneration. In each case, the nerve, shown at the *bottom* of each micrograph, was backfilled with horseradish peroxidase. **A** 4-week regenerate. The sensory pad is poorly developed and only a few nerve fibers are labeled. The labeled cell bodies are not neurons. The *dark material* at the junction of the nerve and epithelium is eye pigment. **B** 10-week regenerate. The *dark band* at the base of the sensory pad is caused by tissue folding. **C** 24-week regenerate. **D** Intact tentacle. Scale *bar* = 250 μm and applies to all micrographs. (Chase and Kamil 1983)

The cerebral ganglia apparently play a role in tentacle regeneration (Fernandez and Flores 1978; Moffett 1991a; Flores et al. 1992). In *Melampus*, removal of the ipsilateral cerebral ganglion at the time of tentacle excision results in regeneration of only a small tentacle, which possibly represents a proportional response to the projections of axons from the remaining (contralateral) cerebral ganglion.

The penis is occasionally lost and in the prosobranch *Crepidula* its regeneration is dependent upon a factor produced by pedal ganglia of young (male-phase) snails (Le Gall 1983; Le Gall et al. 1983). A penis is not regenerated in the pond snail *Lymnaea stagnalis* (Croll, pers. comm.) or the land snail *Helix* (Jeppesen 1976), but in the salt marsh snail *Melampus*, it regrows and functions in mating within 45 days of excision (Moffett 1993). Regeneration depends on the presence of either of two clusters of neurons which innervate it, but function depends upon both clusters (Moffett 1991b, 1992). The role of the neurons in inducing regeneration may be similar to that of serotonergic innervation of the regenerating tentacle (Flores et al. 1992; Chap. 3).

2.9.3 Axon Regeneration

The general observation that regeneration occurs more rapidly and specifically if a nerve is crushed rather than cut is particularly applicable in gastropods. Blood pressure is responsible for changes in body shape, and nerves accommodate significant extension and contraction due to muscles in the nerve sheath. Consequently, when nerves are cut, rather than crushed, the cut ends recoil, resulting in large gaps within an interior that offers little support or guidance. This often poses an insuperable barrier to regeneration. For instance, reinnervation of the salivary gland in *Helisoma trivolvis* is rapid and specific if the nerve is crushed, but it does not occur at all if the nerve is cut (Bulloch et al. 1980). Similarly, recovery of biting behavior following the cutting or crushing of the cerebrobuccal connective in *Aplysia* is seen in animals with connective crushes but not cuts (Kupfermann 1974; Scott and Kirk 1992; Scott, pers comm).

2.9.4 CNS Regeneration

The small snail *Melampus* regenerates axons across cut central tracts and nerves. The cut cerebral commissure of this snail regenerates, first forming a long, thin, bridge of sheath elements and axons which gradually shortens (Price 1977; Moffett and Ridgway 1988a). This regeneration is associated with restoration of reflex responses of the tentacles (Moffett and Snyder 1985). Regeneration also occurs following the removal of a cerebral ganglion (Price 1977; Moffett and Ridgway 1988a,b). Function is restored by tracts that grow through the site of the missing ganglion. In some animals, the neuropilar enlargement gains neurons and a ganglion with the cytoarchitecture of the normal ganglion develops (Moffett and Austin (1982); Moffett and Ridgway 1988a,b). Regeneration of the tentacle ganglion, described above, is also an example of CNS regeneration.

2.10 Echinodermata: The Spiny Deuterostomes

The echinoderms are deuterostome invertebrates, which places them closer to the vertebrates than the other invertebrates discussed so far. It is a characteristic of these animals that they can autotomize body parts and regenerate rapidly

(Emson and Wilkie 1980). A study which mapped neurotransmitter immunoreactivity in the nervous system of the starfish *Asterias* has provided information on the concentration of neurons at a site of arm regeneration (Elphick et al. 1989). Sea cucumbers can eviscerate in response to disturbances or noxious conditions and subsequently regenerate their internal organs. Regeneration of the associated enteric nervous system occurs in parallel with the intestinal repair and is virtually identical to the noneviscerated control (Garcia-Arraras et al. 1994).

Brittle stars get their name from the ease with which they can autotomize part of an arm. Autotomy of starfish arms from the central disc occurs readily; only one starfish, *Linckia*, has been shown to be able to regenerate an entire animal from a single arm (Brusca and Brusca 1990). The role of the nervous system, and particularly of catecholamines, in the regeneration of the arm tip has been examined (Huet 1975). The feather star, a crinoid, has been studied with respect to the role of transmitters (serotonin and dopamine) as organizers of the regeneration (Candia-Carnevali et al. 1989). Sea urchin development has been studied with respect to the role of neurotransmitters (Lauder 1993); serotonin plays a role in gastrulation and migration of the mesenchyme (Gustafson and Toneby 1970).

2.11 Chordata: Our Own Phylum

The invertebrate members of this phylum, the Protochordata, include the Urochordata (ascidians, sea squirts, tunicates) and the Cephalochordata (*Amphioxus*). Particular interest in these invertebrates arises out of their phylogenetic closeness to vertebrates. The immunohistochemical studies of Uemura et al. (1994) on the nervous system of *Amphioxus* are aimed at determining the origin of brain structures associated with neuroendocrine functions in vertebrates. The infolding of the neural plate of urochordates is similar to the vertebrate pattern, and the ascidian embryo is being studied as a prototype of vertebrate embryogenesis (Okamura et al. 1993). Studies of neural anatomy and neurotransmitter immunoreactivity also form the basis of fruitful comparisons with vertebrates and the interpretation of experimental manipulations (Koyama and Kusunoki 1993; Pestarino 1993; Pestarino et al. 1993).

The fact that the central nervous system of tunicates has excellent powers of regeneration has been recognized for many years (Schultze 1899) and mechanisms to account for the origin of new ganglion cells have been proposed (Lender and Bouchard-Madrelle 1964). The simple brain in the sessile adult tunicate bears little resemblence to the brain in the free-swimming tadpole larva, but the entire adult CNS of *Ciona intestinalis* can be regenerated (Bollner 1989; Bollner et al. 1992). Antibodies to peptide neurotransmitters reveal that most neuron categories reappear within 35 days (Fig. 2.14; Bollner et al. 1993). The origin of the new neurons has been followed with a combination of neuron-specific anti-

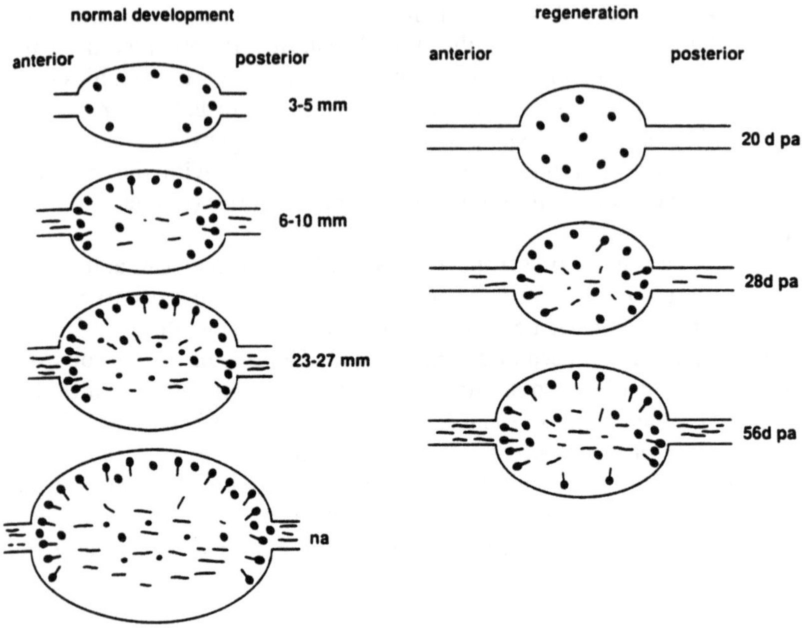

Fig. 2.14. Schematic summary of γ amino butyric acid-like (GABA-li) immunoreactive staining pattern during both normal postmetamorphic development and regeneration. Anterior is to the *left* and dorsal is to the *top*. The earliest stages in both cases are characterized by a lack of immunoreactive fibers, but whereas the immunoreactive cell bodies in normal development are localized mainly in the cortical region, regenerated GABA-li immunoreactive cells are scattered throughout the ganglion. At the 6–10 mm stage in normal development and after 28 days of regeneration, cells with immunoreactive fibers are detected. The 56 day regenerates display more immunoreactive cell bodies and fibers but the GABA cells are distributed differently from the pattern seen in the normal adult (*na*). More cell bodies are found ventrally in the ganglion. The juvenile 23–27 mm stage shows the same pattern as the sexually mature adult. (Bollner et al. 1993)

bodies and bromodeoxyuridine (BRDU) localization (Bollner et al. 1995). These approaches suggest that the new ganglion develops from two sources: mitotic activity in an epithelial tube that has a similar origin to the nervous system, and transformation of preexisting cells, probably blood cells, into one of the types of neurons in the new ganglion.

The phylogenetic closeness of tunicates to vertebrates and their extraordinary capacity for CNS regeneration suggest that additional studies may yield information directly applicable to vertebrate regeneration.

Chapter 3

Early Responses to Neural Injury

3.1 Introduction

Recovery from injury involves the rapid initiation of processes which limit damage, and then the alteration of synthesis and possibly mitosis to rebuild disrupted structure. The signals that activate and coordinate the process of repair are beginning to be appreciated. Greater understanding will allow us to manipulate conditions so as to limit damage and promote recovery in situations where regeneration often fails.

The aime of this chapter is to examine the nature of the responses that lead to successful regeneration. The processes involved in membrane repair, the activation of altered synthetic patterns to support growth, and the roles played by nonneural cells in supporting outgrowth have been clarified in various invertebrate preparations. Conditions under which inadequate repair responses are initiated or the responses of injured neurons are blocked give us insights into the factors crucial to successful regeneration. In addition to providing insights into responses common to most nervous systems following injury, invertebrate neurons offer some novel responses to neural injury, including the formation of electrical connections or fusion of severed axon segments.

3.2 Immediate Responses of Injured Neurons

3.2.1 Is the Injury Discharge a Cry for Help?

The wail of high frequency action potentials given off by an injured neuron (as heard over the audio monitor) is a signal that is difficult for an electrophysiologist to forget. The question I wish to pose is whether a prolonged discharge of action potentials which is significantly different from a cell's normal firing pattern is potentially a signal to the soma, follower cells, or even presynaptic cells, that a neuron has been injured. Such information might be spread by release of neuromodulators such as serotonin, which is released by interneurons upon rapid, prolonged activity in *Aplysia* sensory cells (Glanzman et al. 1989). Serotonin promotes growth through its effect on cAMP pathways that lead to gene activation (Walters and Ambron 1995).

The activation of immediate early genes by neuron depolarization (Bartel et al. 1989; Sheng and Greenberg 1991) and the expression of these genes within 60 min after axotomy suggest that membrane depolarization can be read as an injury signal which second messengers communicate to the nucleus (Leah et al. 1991). Recognition that injury barrages or prolonged depolarization in nociceptive or other neurons may alter circuitry and lead to postoperative (neuropathic) pain has prompted proposals that the activity be preemptively blocked during human surgery (Cervero and Laird 1991).

3.2.2 Basis of an Injury Discharge

Depolarization, reflecting invasion of positively charged extracellular ions, is likely to occur in all neurons experiencing axotomy. Membrane properties at the injury site are not necessarily able to support initiation of action potentials, but in some neurons, axotomy elicits a high-frequency barrage of action potentials that can continue for many minutes.

For instance, an injury barrage is observed following axotomy of a crayfish motoneuron, anal motoneuron L (AML), at all sites in the abdominal portion of the nerve cord through which the axon runs (Fig. 3.1; Muramoto 1993). Evidence that the firing is not a response to synaptic input to the ALM neuron was provided by more proximal injuries and by ion substitutions (Ca^{2+} and Mg^{2+}). These observations also suggest that the major depolarizing ion must be Na^{+}. The interspike interval initially increases exponentially, levels off within 30–60 s, and declines over a period of minutes. Response latency, duration, and spike frequency are correlated with the site of axotomy and are not dependent upon an intact cell body.

3.2.3 Effects of Depolarization and Ca^{2+} Entry

Injured neurons may not respond with a barrage of action potentials if the injured region does not possess voltage gated membrane channels and the depolarization that reaches the spike-initiating site is subthreshold. No change is seen in the soma in buccal neuron 5 of *Helisoma* when the distance to the injury site is more than three times the space constant. However, the duration of the electrical changes can be followed in the soma when the event is closer. The soma membrane potential maximally depolarizes within 1 min and gradually repolarizes over several hours (Berdan et al. 1993).

Of the ions that enter the cut axon, Ca^{2+} appears to be most intimately associated with the cell's subsequent responses. For instance, the dependence upon Ca^{2+} entry in membrane sealing in cockroach giant axons was revealed by reducing its availability in the external medium and injecting it intracellularly (Yawo and Kuno 1983,1985); eliminating all Ca^{2+} in the external medium at the time of axon crush blocks injury-induced hyperexcitability (Walters and Ambron 1995).

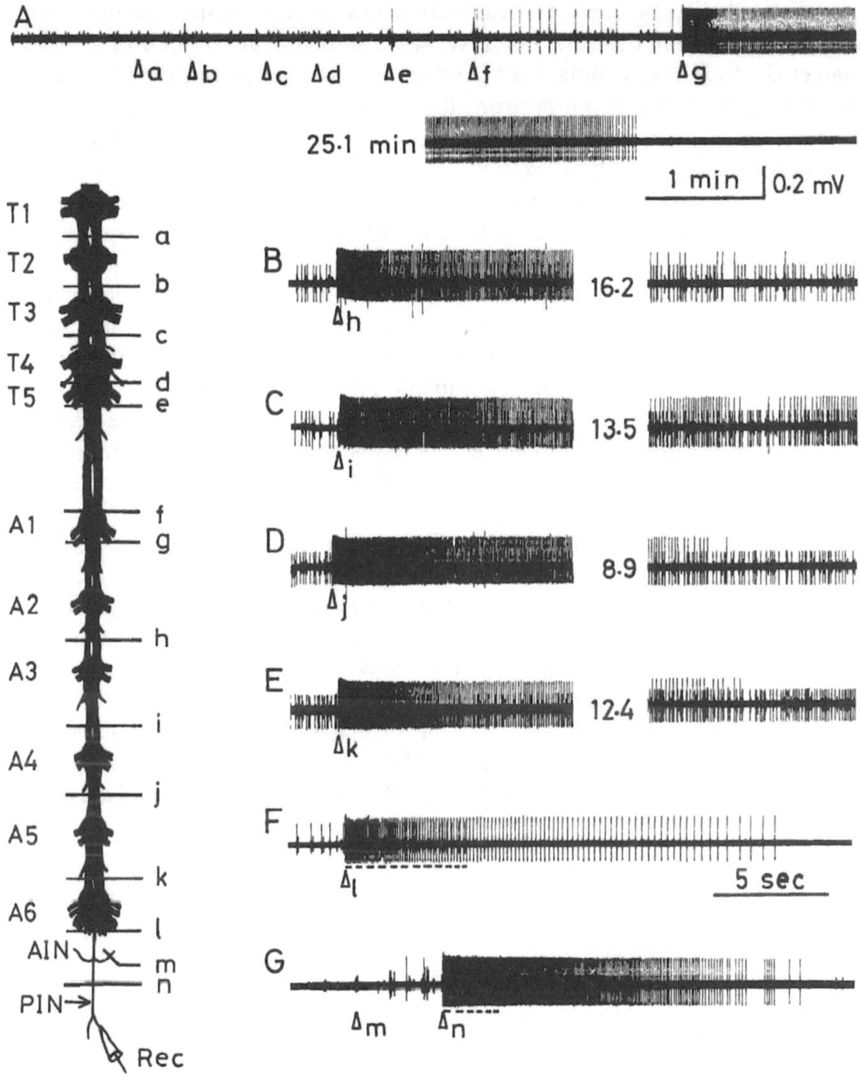

Fig. 3.1. Response of anal motoneuron L to transection (axotomy) at various positions in the isolated nerve cord preparation. The diagram on the *left* shows the nerve cord. The soma of L is in *A6* and an axon runs forward to *A1* and posterior in the posterior intestinal nerve (*PIN*), where recordings are made. Transection positions (*a-n*), and the responses recorded in *PIN* corresponding to each site of transection are indicated with *triangles* associated with the *small letters* on recordings **A-G**. Long-lasting firing was elicited by all cuts that transected the axon, and by none that only activated synaptic inputs (*a-f*). In **B-E**, the initial and final portions of the record are shown, with the time lapse (min) between the two segments of the recording given between recording segments. The long lasting firing was even produced when the segment of the axon transected was no longer connected to the soma (**G**). (Muramoto 1993)

The Ca^{2+} distribution after transection has been measured using the Ca^{2+} sensitive fluorescent dye fura 2 in larval sea lamprey Mauthner neurons (Strautman et al. 1990). Intracellular Ca^{2+} concentration at the cut end rises to 14 times the resting level due to entry from the extracellular environment; a field of elevated Ca^{2+} detected farther from the cut may result from voltage-gated Ca^{2+} channels opened by depolarization or second messenger systems and also release of internal Ca^{2+} stores. Ca^{2+} concentration in the end of the axon returns to near normal within 24 hours as the front of elevated Ca^{2+} travels up the axon at a rate of 0.5 mm/h (Fig. 3.2).

Intracellular Ca^{2+} elevation at the injury site may produce an electrical field that drives the front of positivity toward the soma. The strength of this signal can be manipulated by changing the external Ca^{2+} or the electric field. Application of the Ca^{2+} ionophore A-23187 to the site of injury in fish optic nerve increases the rate of regeneration (Meiri and Grafstein 1984). Imposed electric fields promote regeneration in vertebrate limbs, and in the earthworm, electric fields and focused laser pulses have been used to promote reconnection of axonal stumps (Sect. 3.4.4).

As the front of Ca^{2+} moves within the axon, it will disrupt the cytoskeleton and affect many regulatory enzymes. Considering the high levels of intracellular Ca^{2+} measured, the reestablishment of structural and functional properties in all but the very tip of the lamprey axons was unexpected (Strautman et al. 1990). The high concentrations of Ca^{2+} and associated depolarization may down-regulate existing Ca^{2+} channels, restoring normal intracellular Ca^{2+} levels. Depolarization suppresses both the high-voltage-activated and low-voltage-activated Ca^{2+} currents of an identified *Helisoma* neuron (Berdan et al. 1993). Effects of Ca^{2+} influx on the growth cone (Bedlack et al. 1992) are discussed in Chapter 4.

3.2.4 Repairing the Axon Membrane

If a neuron is to avoid death following axotomy, closing the hole in the cell membrane is a first priority. A vesicular mechanism similar to neurotransmitter release has been described in a number of cell types (Steinhardt et al. 1994). How effective the cell's responses will be is partly determined by the axon diameter, with the largest axons exceeding the size that can be easily blocked by the vesicles that are generated. For instance, the squid giant axon is never effectively closed (Krause et al. 1994), the membrane of earthworm medial giant neurons reseals within 30 min (Todorov et al. 1992) to 1 h (Krause et al. 1994). The membrane potential recorded in the soma of the cockroach giant axon returns to resting level 5–40 min after transection (Meiri et al. 1981). Yawo and Kuno (1985) found that cockroach giant axons form an electrical and dye seal 5–30 min after transection (Fig. 3.3) if temperature and extracellular Ca^{2+} concentration are within the normal range.

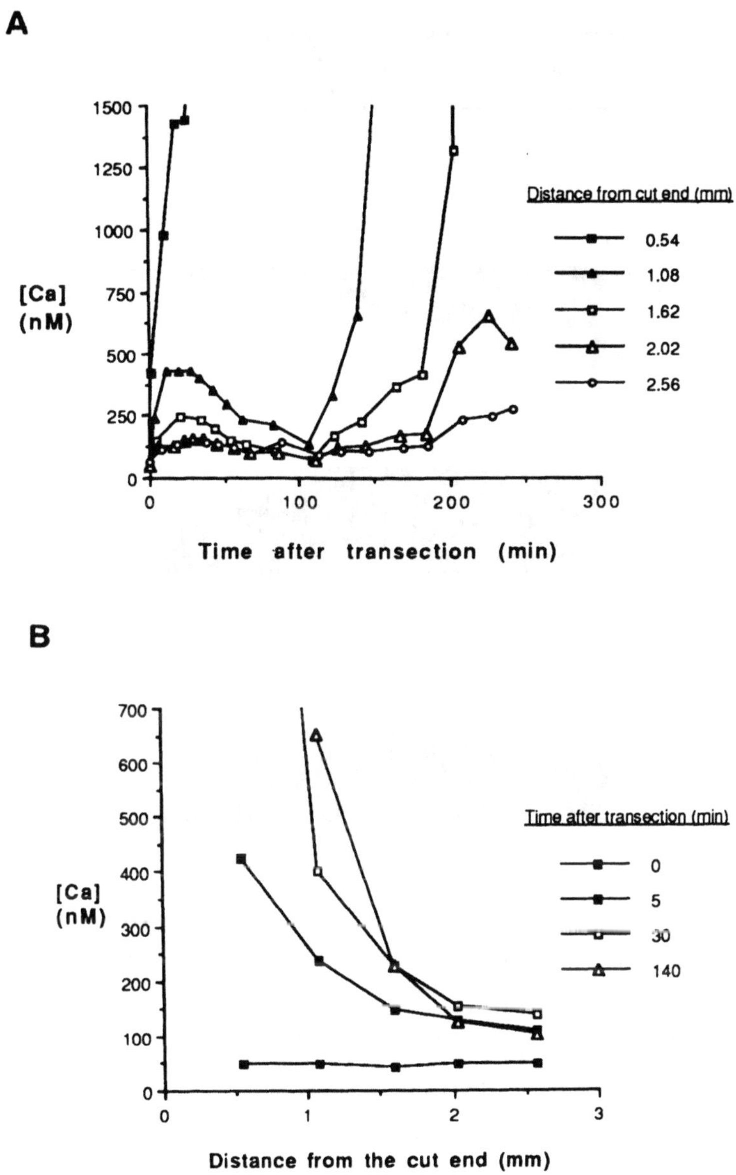

Fig. 3.2. Longer-term changes in $[Ca^{2+}]$. Observations on the axon of the Mauthner cell of larval sea lampreys. The data points are representative of 4 separate experiments. **A** $[Ca^{2+}]$, as a function of time after transection. Each *line* connects values measured at the indicated distance from the cut end. After 4 h, the fura 2 concentration became too low to give reliable values. **B** Replot of data in A showing the change in the $[Ca^{2+}]$ gradient over time (Strautman et al. 1990)

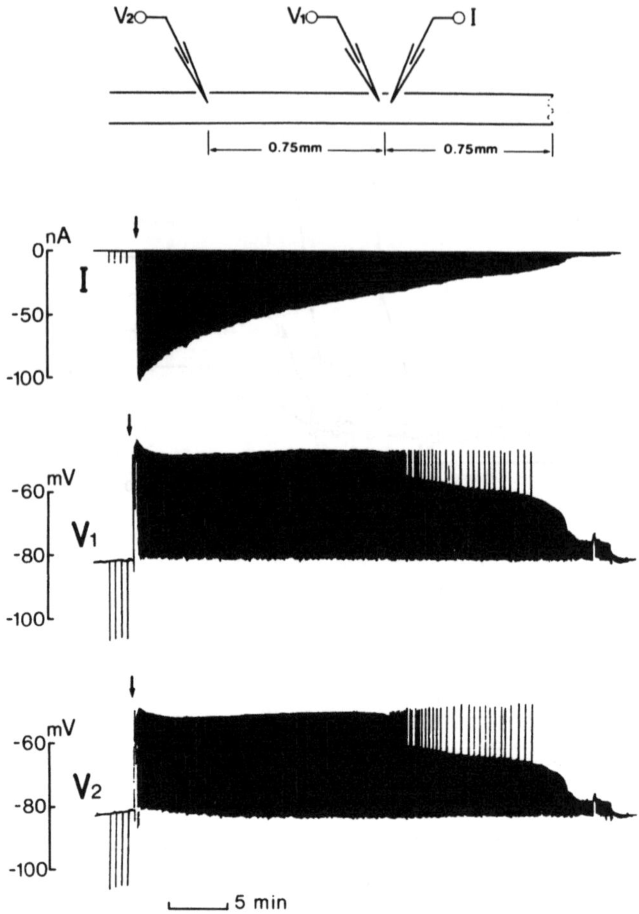

Fig. 3.3. The injury current profile observed under intermittent voltage clamp following axonal transection. *Inset* A schematic diagram illustrating the positions of two recording electrodes (V_1, V_2) and an electrode (*I*) used for passing clamp currents. The axon was manually clamped at about -80 mV for 200 ms every 3 s. **I** Inward injury currents observed following the transection. **V1** Changes in the membrane potential at the site of the intermittent voltage clamp. **V2** Same as **V1**, but at 0.75 mm from the site at which the clamp currents were applied. Axons with a relatively large space constant (>9 mm) were used for this series of experiments, and delayed rectification was minimized by TEA (50 mm) and 4-aminopyridine (3 μg/ml). After the axonal transection, the length constant decreased to about 40–50% of the original value. The degree of space clamp was, however, satisfactory at 0.75 mm from the clamp current source even after the transection. Therefore, the injured site (0.75 mm) was assumed to be under voltage clamp. Note action potentials elicited by the release of the clamp current pulse in **V1** and **V2**. (Yawo and Kuno 1985)

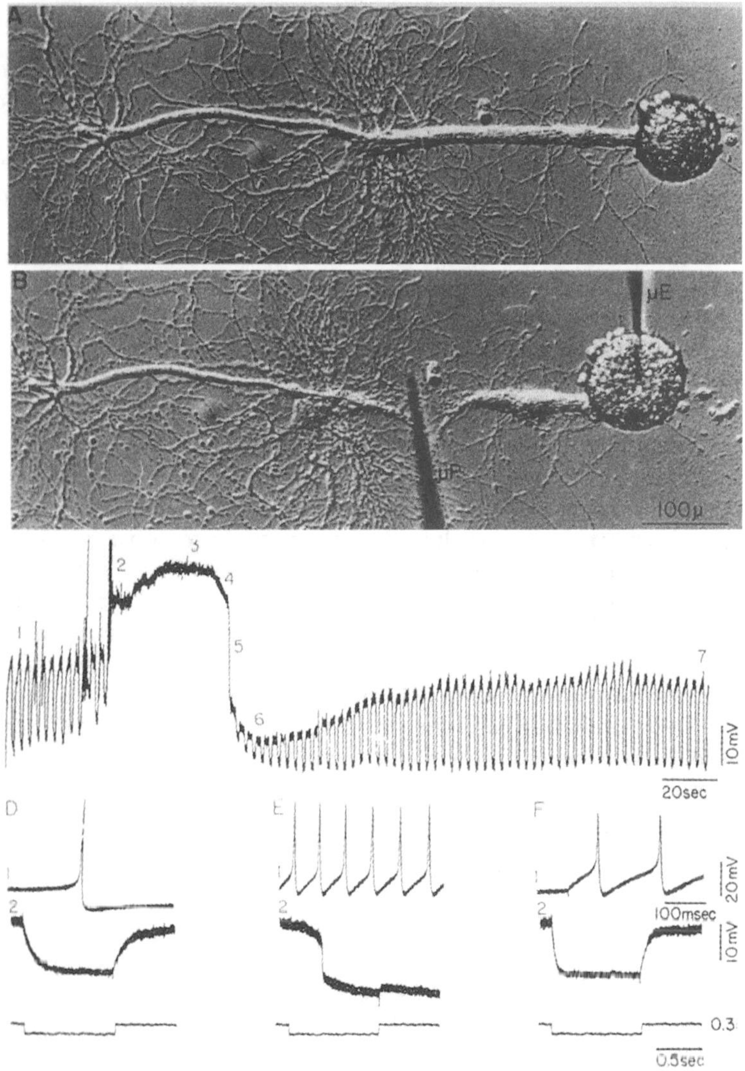

Fig. 3.4. Changes in resting potential and input resistance during axotomy and throughout the recovery process. Metacerebral neuron (MCn), **A** prior- and **B** post-axonal transection by a micropipette (μ*P*). Current injection and voltage recording were done by an intracellular microelectrode (μE) inserted into the cell body. **C-F** A continuous record of the transmembrane potential and input resistance during axotomy and throughout the recovery process. The input resistance was monitored by measuring the voltage drop in response to injection of constant rectangular current pulses of 0.35 nA. Prior to axotomy, the input resistance was 54 MO (*C1*, *D2*), the time constant was 135 ms (*D2*) and the cell fired an action potential in response to a depolarizing pulse (*D1*). Axonal transection produced repetitive firing (*C2*, *E1*); the membrane depolarized by 28 mV and the input resistance was largely decreased (*C3*). After about 40 s the membrane potential recovered, first slowly (*C4*) and then at an accelerated rate (*C5*, *E2*). The membrane potential undershot the resting potential by -21 mV (*C6*), and thereafter the potential gradually recovered (*C7*). Simultaneously, the input resistance increased and slightly exceeded its control value (61 MO)(*F2*). Nevertheless the time constant became shorter than in the control (51 ms)(*F2*), indicating the neuronal membrane had a higher conductance. During this period, the cell regained its ability to fire action potentials (*F1*). (Spira et al. 1993)

Gastropod neuron responses to axotomy have been studied both in vivo and in vitro. *Helisoma* buccal neuron 5 forms a barrier to diffusion of carboxy-fluorescein at the site of axotomy within 15 min of connective crush (Berdan et al. 1993). (Compare with the data for cultured septal neurons of the rat, which seal within 10 min of transection: Xie and Barrett 1991). In the cultured metacerebral giant neuron of *Aplysia*, resealing after axotomy occurs within 10–260 s, as evidenced by recovery of membrane potential and input resistence in both the proximal and distal cut ends of the metacerebral cell (Fig. 3.4). The injury discharge can continue for the duration of the depolarization (Spira et al. 1993). Evidence of such rapid sealing was seen occasionally in recordings from *Helisoma* neurons (Berdan et al. 1993).

Axons may become partly or completely sealed through a combination of constriction and blockage by coalescing membrane vesicles, as described for squid and earthworm giant axons (Gallant 1988; Fishman et al. 1990; Krause et al. 1994). Constriction involves softening due to depolymerization of axoplasmic cytoskeleton and elution of cytoplasm in response to extracellular ions (Fig. 3.5). Calcium is important in inducing constriction, but Cl^- and Mg^{2+} also contribute to the response. Elevated external Ca^{2+} has been linked to destabilization of cytoskeletal components through activation of endogenous proteases; other treatments that tend to destabilize microtubules and neurofilaments also promote membrane sealing (Xie and Barrett 1991).

Vesiculation begins in the squid axon within 1 min of axon transection to produce axosomes, large vesicles composed of membranes from intracellular sources (Fishman et al. 1990). Some vesicles flow out of the cut end of the axon and others contribute to plugging the narrowed opening (Fig. 3.6).

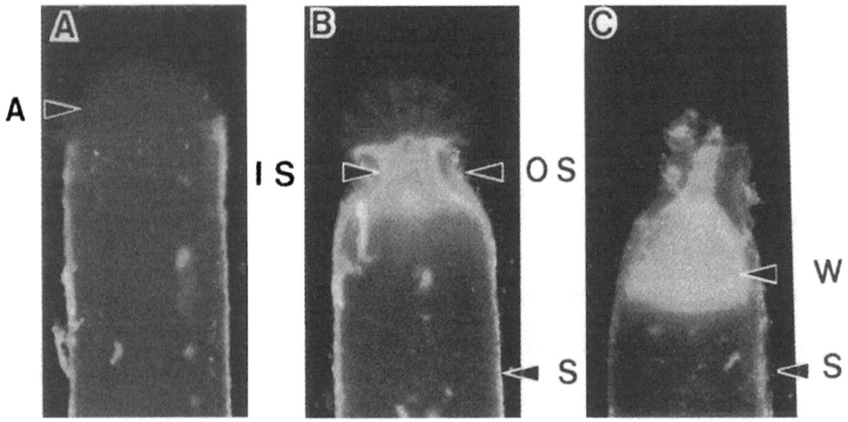

Fig. 3.5. Constriction and closure of the squid giant axon after transection. The axon was cleaned, placed in the standard external medium, and photographed. **A** Several seconds after; **B** 5 min after; and **C** 100 min after transecting the axon with a pair of microdissection scissors. The sheath (*S*) surrounding the axon split into an inner sheath (*IS*) and an outer sheath (*OS*) at the site of constriction. Axoplasm (*A*) began to bulge out of the cut end after transection, and the cut end began to turn white (*W*) with time. (Gallant 1988)

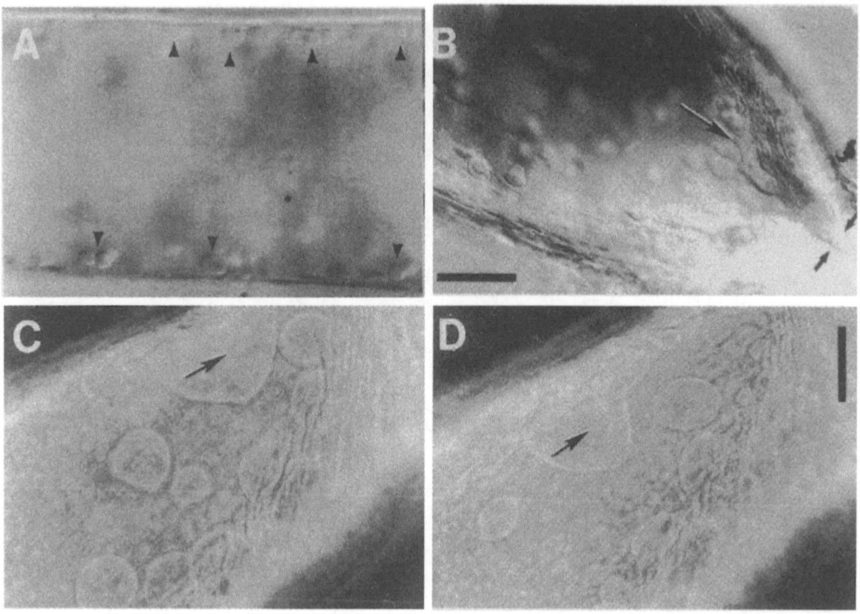

Fig. 3.6. Axosomal movement and fusion in a squid giant axon transected in ASW. **A** Generation of vesicles (axosomes) along the axolemma. **B** Axosomal accumulation and occlusion at a constricted cut end of an axon. **C** A large irregularly shaped axosome '*A*' (*arrow*) before fusion. **D** 7 min later the same axosome has changed shape and increased in area, apparently due to fusion, and has moved toward the cut end (not shown) in the direction of the *arrow*. The rate of movement was calculated to be about 20 μm/min, which is suggestive of a translocation mechanism. Note other axosomal size increases and movements during the elapsed time from C to D. Phase contrast. Scale bar = 00 μm for A, C and D, 50 μm for B (A, Fishman, unpubl; B-D, Fishman et al. 1990)

The medial giant axon of the earthworm is plugged by injury-induced vesicles that form a tightly-packed plug at the constricted end (Krause et al. 1994; Fig. 3.7). A similar process, reflected by changes in membrane resistance and the appearance of a partition-like structure near the cut end, may occur in cockroach giant neurons. The use of blockers suggests that activation by Ca^{2+} of phopholipase A2 is important in cockroach membrane resealing (Yawo and Kuno 1985).

In contrast to the observation that vesicles and constriction together constitute the closure mechanism for squid and earthworm axons, fusion of the axon membrane is the only event that is correlated with electrical evidence of a seal in an *Aplysia* neuron (Spira et al. 1993). The metacerebral giant neuron of *Aplysia* was axotomized in vitro with intracellular electrodes in place and provision for rapid fixation. Under these conditions, the ultrastructural condition at the time of electrical evidence of seal formation could be determined (Spira et al. 1993). Immediately after axotomy and in the tens to hundreds of seconds before repolarization, a small transmembrane resistance was recorded which is attri-

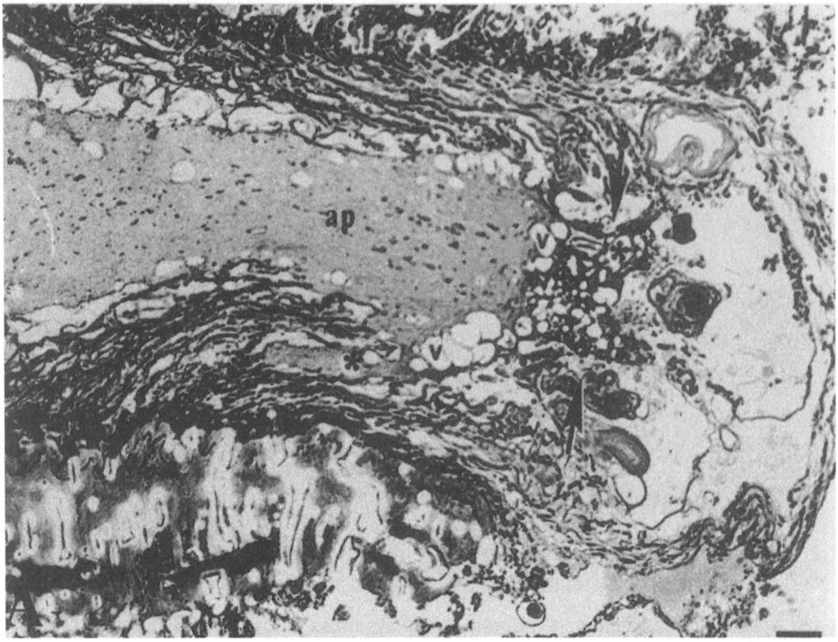

Fig. 3.7. Light micrograph of a typical earthworm medial giant axon fixed 1 h posttransection. The axon is longitudinally sectioned. Note that the cut end is slightly constricted, and large intraaxonal vesicles can be seen along the axolemma and accumulated at the cut end; *ap* axoplasm. Scale *bar* = 10 μm. (Krause et al. 1994)

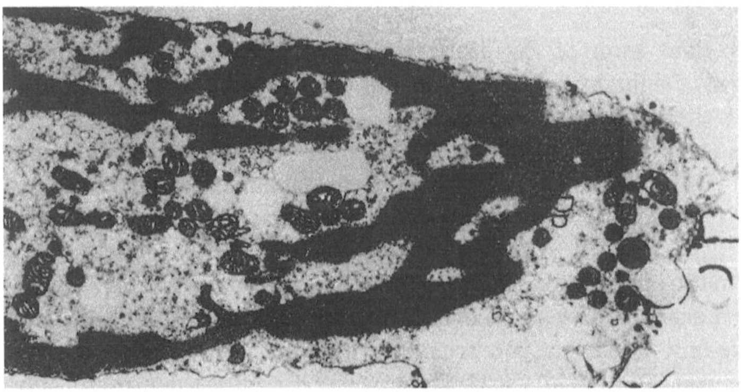

Fig. 3.8. Electron micrograph of the proximal segment of a transected metacerebral neuron of *Aplysia* after the rapid recovery phase. At the tip of the transected neuron, electron dense deposits are aggregated forming a cap-like structure, the cytoplasm is made of amorphous material, no cytoskeletal elements are seen but there are normal-looking mitochondria, which are disoriented. Serial longitudinal sections revealed a continous plasma membrane over the tip of the axon. (Spira et al. 1993)

butable to the cytoplasm, because it is linearly distributed along the axon core and not localized at the electron dense deposits that accumulate at the axon tip before a membrane seal is detected. Recovery of the resting potential and membrane resistance is correlated with the membrane barrier which forms after cytoplasm is extruded and the axolema fuses across the narrowed tip (Fig. 3.8).

Information on the nature of the membrane that seals the end of the axon has been inferred from membrane selectivity properties. The repair sequence begins in the nonselective state (no intact membrane), progresses to a resting selectivity that differs from the control and supports a Ca^{2+}-dependent action potential, and finally restores the normal membrane properties (Meiri et al. 1981).

In gastropod neurons, considerable retraction occurs as both axon segments seal, forming retraction bulbs. Injury does not have to be adjacent to the soma for measurable retraction to occur (Croll and Baker 1990), but if axotomy is close to the soma, complete resorption of the neurite may occur (Murphy and Kater 1980; Bulloch and Kater 1981; Hadley et al. 1982).

3.2.5 Changes in Excitability Following Axotomy

Axotomy-induced alterations in electrophysiological properties are often seen in vertebrate and invertebrate neurons (Titmus and Faber 1990; Matzner and Devor 1994). Some leech neurons exhibit no changes in electrical properties during the period of sprouting (Bannatyne et al. 1989), but others exhibit increased excitability, as noted below. Larval insect sensory neurons revert to a less differentiated state after axotomy, losing the capacity to conduct action potentials, and regaining it only after contacting their target (Chiba and Murphey 1991). Axotomized motoneurons of the locust exhibit increased excitability following axotomy or colchicine treatment that is the result of an increase in the number of voltage sensitive Na^+ channels in the soma (Goodman and Heitler 1979). Such increases in excitability may increase the growth potential of neurons, as a result of second messenger systems activated by Ca^{2+} entry.

The increased sensitivity may be an epiphenomenon consequent to the rapid production of replacement membrane for the axon repair, as suggested by some of the examples below, or it may reflect activation of pathways that record a memory of the injury and provide for altered behavior (see Sect. 3.2.6). The distinction between these two mechanisms may not be easy to determine in some systems.

As synthesis to repair one region of membrane is increased, incorporation of inappropriate channels into uninjured regions of membrane apparently occurs, introducing changes in excitability. Membrane components are not the only repair elements that are misappropriated. The normal segregation of other subcellular specializations may be lost in the rapid repair following injury. For instance, in lamprey neurons, neurofilaments that are normally confined to axons are incorporated into the dendrites following axotomy close to the soma (Hall et al. 1989).

Upheaval in cytoskeletal pathways to peripheral sites may be responsible for anomalous channel distribution. If the cytoskeletal tracts in an axon are broken by axon compression that does not disrupt the plasma membrane, the fast axoplasmic transport system is disrupted, temporarily blocking vesicle distribution (Gallant 1992). In cultured neurons that "ball up" after transection of the axon close to the soma, the microtubules have no parallel orientation, and neurites are extended in a radial pattern rather than being restricted to projections from the end of the axon (Benbassat and Spira 1993). The addition of membrane channels in a radial pattern may explain the observation that when the neurons that fuse during development to produce the squid giant axon are cultured individually, they form round cells, and excitable sodium channels are introduced into the normally inexcitable soma (Gilly and Brismar 1989).

The hyperexcitability seen in mammalian afferent axons at a neuroma is dependent upon the accumulation of voltage-sensitive sodium channels not normally present in the afferent axonal membrane (Matzner and Devor 1994). Chromatolysed cat motoneurons exhibit spike-like responses in dendrites that elevate the responses to individual inputs (Kuno and Llinas 1970; Perry et al. 1993).

The timing of alterations in electrogenicity following axotomy is correlated with cytological evidence of metabolic changes in the soma (Pitman et al. 1972; see Sect. 3.3.3). Blockers of protein synthesis indicate that new Na^+ channels synthesized within 24–78 h in response to axotomy or deafferentation are incorporated into the repair membrane in the insect dorsal unpaired midline (DUM) neurons (Tribut et al. 1991). Normally phasic, relatively inexcitable neurons of arthropods exhibit increases in excitability 1.5–16 days after axotomy (crayfish motor neurons) or within 6–48 hours (cricket medial giant interneurons) (Kuwada and Wine 1981; Roederer and Cohen 1983). In locust neurons, increased excitability of the soma following axotomy is attributable to the addition of voltage-sensitive Na^+ channels (Goodman and Heitler 1979; Kuwada 1981). A similar response is seen in anterior pagoda neurons of the leech, which normally have a nonspiking soma but exhibit progressive incorporation of membrane patches that support elevated excitability close to the soma after injury (Pellegrino et al. 1984). Even electrical connections are affected by axotomy: decreases in junctional resistance of the electrotonic synapse between VD1 and RPD2 of *Lymnaea* occur so rapidly (within 3 h) after distant axotomy that soma-driven changes in the properties of incorporated membrane are not likely, so the authors suggest that alterations in existing channels or sprouting that alters contact area may be responsible (Pilkington and Benjamin 1986).

3.2.6 Excitability Changes: Was Injury the First Teacher?

Evidence for changes in circuit function in axotomized neurons has revealed marked similarities to changes brought about by learning paradigms in *Aplysia* neurons (Walters et al. 1991). As in learning, sensory neurons exhibit increased excitability and improved synaptic efficacy following axotomy (Fig. 3.9), and the changes are initiated following transmission of a retrograde signal by reverse axoplasmic transport.

Tetanic stimulation, which resembles the injury discharge, results in long-term potentiation (Fazeli 1992). The injury discharge can therefore have a direct

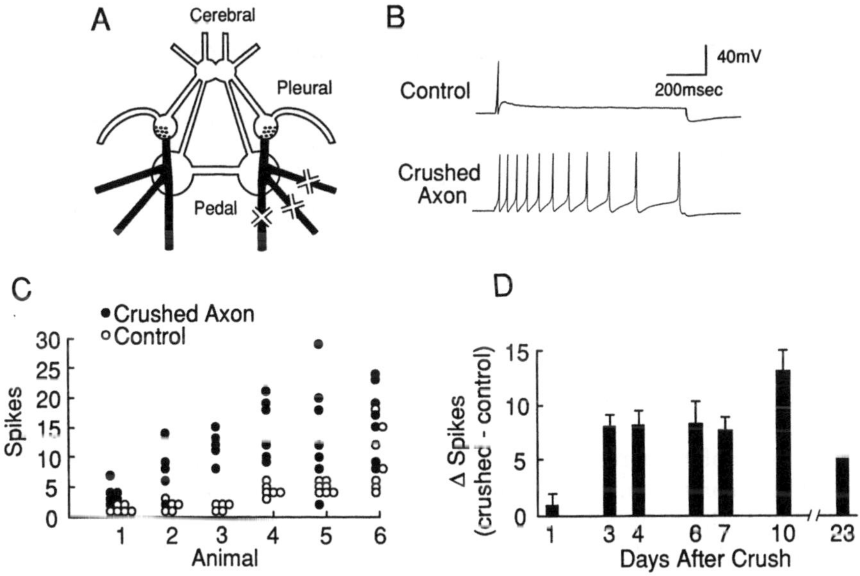

Fig. 3.9. Long-term alteration of action potential accomodation induced by nerve crush. **A** Sensory neuron somata (*dots*) are located in each pleural ganglion and project their axons (*dark lines*) through the ipsilateral pedal ganglion into pedal nerves. The *X*'s indicate crush of pedal nerves 15–30 mm from pleural somata. In some experiments the connective between the pleural and pedal ganglia was crushed on one side. **B** Action potential (*spike*) accommodating in a pair of sensory neurons during 1-s depolarizing test pulses 4 days after unilateral pedal nerve crush. Accomodation is defined as a decline in spike frequency during the pulse. **C** Number of action potentials evoked by 1-s test pulses 4 days after unilateral pedal nerve crush. Each point (*filled circles*, crushed axon; *open circles*, control) represents the response of a single sensory neuron. **D** Mean (+ SEM) differences between the number of spikes evoked by the 1 s test pulses in sensory clusters subjected to axon crush and those evoked in contralateral control clusters. The number of spikes evoked in each cell within a cluster was averaged, yielding a pair of spike scores for each animal. These scores were then averaged across each group of animals (one group per test interval), and the mean difference was plotted at each test interval. (Walters et al. 1991)

role in inducing changes in excitability and the growth potential of the cell (Walters and Ambron 1995). The effects of action potential discharge and the effects of retrogradely conducted signals in response to injury probably converge at the level of gene induction to produce the total response activated by injury. When possible contributions of an injury discharge per se or central changes in excitability due to release of neuromodulators are minimized by crushing the nerve under conditions that reduce action potential discharge and synaptic transmission, changes in excitability are still seen in the sensory cells that persist 3–23 days following axotomy. The role of the retrogradely transported signal is described in Section 3.3.2.

3.3 Growth Following Axotomy

3.3.1 Initial Outgrowth is Independent of Soma Responses

After the gaping hole in the axon is closed and before changes in synthesis and transport from the soma can be reflected at the site of injury, neurites begin to grow. Sprouting is seen at both the proximal (cell attached) and the distal (anucleate) ends of axons. This attests to the capacity of the cut ends to incorporate membrane, perhaps as an extension of the healing process that seals the stumps (Kennedy and Bittner 1974; Govind et al. 1992). It also attests to the relative (and temporary) independence of the far-flung neuronal projections from processes controlled within the soma.

Sprouting is seen within minutes in *Aplysia* neurons (Spira et al. 1993). Berdan and Easaw (1992) found that the distance of the lesion from the soma did not influence the latent period to sprouting, an observation that is consistent with studies of vertebrate axon sprouting (Gutmann et al. 1941). This is in contrast to the correlation of distance separating lesion and soma and the delay in initiation of the soma reaction.

Nine h after axotomy, 31% of *Helisoma* buccal neurons 5 (B5) exhibit outgrowth, and 100% sprout within 13 h of axotomy in organ culture (Jones and Bulloch 1988; Berdan and Easaw 1992). Initiation of sprouting is independent of protein synthesis blockers and is not accelerated by temperature elevation. Various channel agonists and blockers reveal that decreasing intracellular Ca^{2+} stimulates sprouting and elevating Ca^{2+} suppresses it. Forskolin, an activator of adenylate cyclase, also inhibits sprouting. These observations cannot necessarily be generalized to all neurons in *Helisoma*, as neurons exhibit individual responses to the same signals and the "calcium set point hypothesis" suggests that an increase or decrease in intracellular Ca^{2+} is read relative to the preexisting Ca^{2+} level and the cell's physiological state (Mills and Kater 1990).

Some protein synthesis may occur in growth cones, which in many species possess polyribosomes (Davis et al. 1992). The isolated dendritic growth cones of the snail *Helisoma trivolvis* show a high capacity for ribosomal-dependent translation, whereas axonal growth cones do not (Davis et al. 1992). Long-term survival of distal segments of crustacean motoneurons is dependent upon the incorporation of glial nuclei (Parnas et al. 1991; Sect. 3.4)

3.3.2 Retrograde Signals

In many invertebrates and cold-blooded vertebrates, growth cone elongation may start as a local response that grades imperceptibly into the soma-directed response as material synthesized in the soma reaches the growth cone. The onset of the soma response is typically correlated with a change in the composition of retrogradely-transported signals. Possible reflections of damage include 1) a positive signal (damage out there), 2) loss of a trophic signal from the target (no more growth factors, etc.) 3) loss of feedback from activity (recycled vesicles), 4) loss of a repressing signal (no more growth inhibitor), or 5) synthesis of a new signal (the situation has changed).

One class of feedback is associated with depolarization and synaptic vesicle release. Recovered membrane and extracellular signals are retrogradely transported in recycled synaptic vesicles. Blocking electrical activity can prevent transmitter release and reduce vesicle membrane recycling. Action potentials blocked by focal application of tetrodotoxin (TTX) to a motor nerve can produce axotomy-like changes that are reversed by stimulating the nerve distal to the block (Czeh et al. 1978). In cat spinal motoneurons, blocking acetylcholine (ACh) release with botulinum toxin leads to axotomy-like changes in excitability in the most strongly affected motoneurons (Pinter et al. 1991).

Interfering with synaptic function several steps earlier in the sequence, by blocking serotonin synthesis or injecting 5,7-dihydroxytryptamine, a depleter of serotonin in gastropods, produces an axotomy-like effect in the soma of serotonergic neurons. It also affects non-serotonergic cells, as evidenced by the sprouting seen in an otherwise intact adult snail nervous system (Baker and Croll 1991; Baker et al. 1993).

Studies in which normal retrograde transport is reversibly interrupted can initiate the soma response in some neurons. Colchicine treatment mimics the effects of axotomy on electrical and cytologic characteristics of cricket and locust neurons, presumably by blocking axoplasmic transport (Pitman et al. 1972; Goodman and Heitler 1979).

The nature of the retrograde signal is not clear in many neurons, but the growth cone component GAP-43, a developmentally-regulated protein, is reexpressed in axotomized cells following loss of the inhibition normally provided by a target-derived repressor (Skene 1992).

If the somata of mammalian motoneurons are deprived of signals delivered by retrograde transport from the target or from the site of axotomy by application of vinblastine to an intermediate site, the soma response is still initiated (Tornqvist et al. 1994). This would argue that more than one factor is operating or that no target-derived or damage-generated signals are necesary. What other signals could the soma be getting? One possibility is that glia may produce a relevant signal. Glia and other nonneural cells proximal to a transport block could respond to axonal injury or the lack of normal axonal transport by synthesizing a new signal that is picked up and retrogradely transported or detected by a membrane-bound receptor system. The question then becomes: how do glia detect injury? An example of an axon-to-glia communication loop that would be interrupted by injury or blockage is the extrasynaptic release of the neurotransmitter glutamine by crayfish giant axons and its uptake by glia (Grossfeld et al. 1994).

Evidence for a positive signal reflecting injury relayed to the soma by retrograde transport has been provided in studies of *Aplysia* sensory neurons (Walters and Ambron 1995). Retrograde transport is required for the hyperexcitability induced in those cells, and blockers of transport block the response. The events in the injury response sequence appear to be the following. A class of molecules that are constitutively transported in an anterograde direction are activated by axon injury (probably consequent to protein kinase activation by Ca^{2+} entry; see also Conner and Smith 1994) to reveal a nuclear localization signal. These altered molecules, called sp97, are picked up by the retrograde transport system and moved up the axon and straight into the nucleus (Walters and Ambron 1995). Retrograde transport is accelerated by a factor of two to three by cAMP, which will normally be produced in the injured cells in response to serotonin (Azhderian et al. 1994). Thus the neuromodulatory consequence of rapid firing would normally speed the arrival of the retrogradely transported injury signal.

3.3.3 Anatomical Correlates of the Soma Response

Anatomical alterations in soma structure in response to axotomy are variously referred to as the axon reaction, the soma reaction or the chromatolytic response (Lieberman 1971). Typical changes observed with the light microscope within 2–3 days of axotomy include reduction in basophilic staining (primarily bound rRNA), swelling of the cell, and movement of the nucleus to an asymmetric position. Axotomized insect neurons swell and develop a perinuclear ring, the opposite of a chromatolysis reaction (Cohen and Jacklet 1965; Byers 1970). Following axotomy, cricket neurons also accumulate vacuoles and clusters of microtubules close to the initial neurite; these changes are also seen following alterations in excitability (see Sect. 3.2.5), and disappear within 10 days (Roederer and Cohen 1983).

Neuron B5 of *Helisoma* undergoes a chromatolysis reaction, i.e., the dispersion of Nissl bodies, which at the ultrastructural level corresponds to a less dense pattern of rough endoplasmic reticulum and associated polysomes. These changes are seen earlier in isolated somata than in organ cultured neurons (Berdan et al. 1990; Fig. 3.10). Young's (1932) study of cephalopod neurons also includes a chromatolysis reaction following axotomy.

Observations of alterations in soma staining properties may be quite useful for correlation with functional adjustments in a particular cell type. However, such gross changes are not found in all cells undergoing a soma response, their significance is difficult to interpret, and they may not be correlated with the capacity or strength of the regeneration response (Schwartz 1987).

3.3.4 Metabolic Adjustments Associated with Regeneration

The changes in metabolic activity brought about by axotomy reflect shifts in the pattern of genetic readout of the cell. The expression of immediate early genes predicts changes in expression of growth factors and structural elements. In general, synthesis of cytoskeletal elements after axotomy would be expected to favor those elements that support growth. Neurofilament disassembly will allow shape changes for debris removal and initiation of growth.

In the vertebrate CNS, responses to axotomy range from increased RNA transcription and protein synthesis to regressive cytological and metabolic changes which presage death (Reis and Ross 1973; Ross et al. 1975). Within 1–3 h of axotomy or deafferentation, those basal ganglion populations that survive exhibit c-fos induction (Weiser et al. 1993; Curran and Morgan 1995). Many conditions increase the expression of c-fos, which is able to bridge the gap between extracellular signals and the alteration of gene function.

An immediate early gene product, CCAAT enhancer binding protein (C/EBP), is activated both in the serotonin induction of *Aplysia* synaptic facilitation and by tissue injury alone (Alberini et al. 1994). This common pathway for facilitation and injury responses supports the impression that injury may have been the first teacher.

Synthesis associated with axonal growth increases at later stages in some neurons. Markers of the growth state, such as the structural component $T\alpha_1$ tubulin (β-tubulin) and growth-related peptides, particularly GAP-43, are transported down the axon at different rates: the GAPs are carried by fast axonal transport, and have rapid effects, whereas the structural components are transported by slow component b of axonal transport (Schwartz 1987; Jacob and McQuarrie 1991). After axonal injury, the rate of slow transport may increase in regenerating axons and slow in those axons that cannot regenerate (McKerracher and Hirscheimer 1992). Lipid transport also may increase (Schwartz 1987).

In mammals, responses to injury or target deprivation involve cell death (apoptosis) in some neurons. Incorporation into DNA of an analog of thymidine, bromodeoxyuridine, indicates that the process involves abortive reentry into the mitotic cell cycle (Herrup and Busser 1994; McCarthy et al. 1994; Rossiter et al. 1994). A similar response with a successful outcome is seen in the jellyfish. Striated muscle cultured in the absence of extracellular matrix or in the presence of a monoclonal antibody to the extracellular matrix exhibited DNA replication and transdifferentiation into neurons or smooth muscle cells (Reber-Muller et al. 1994).

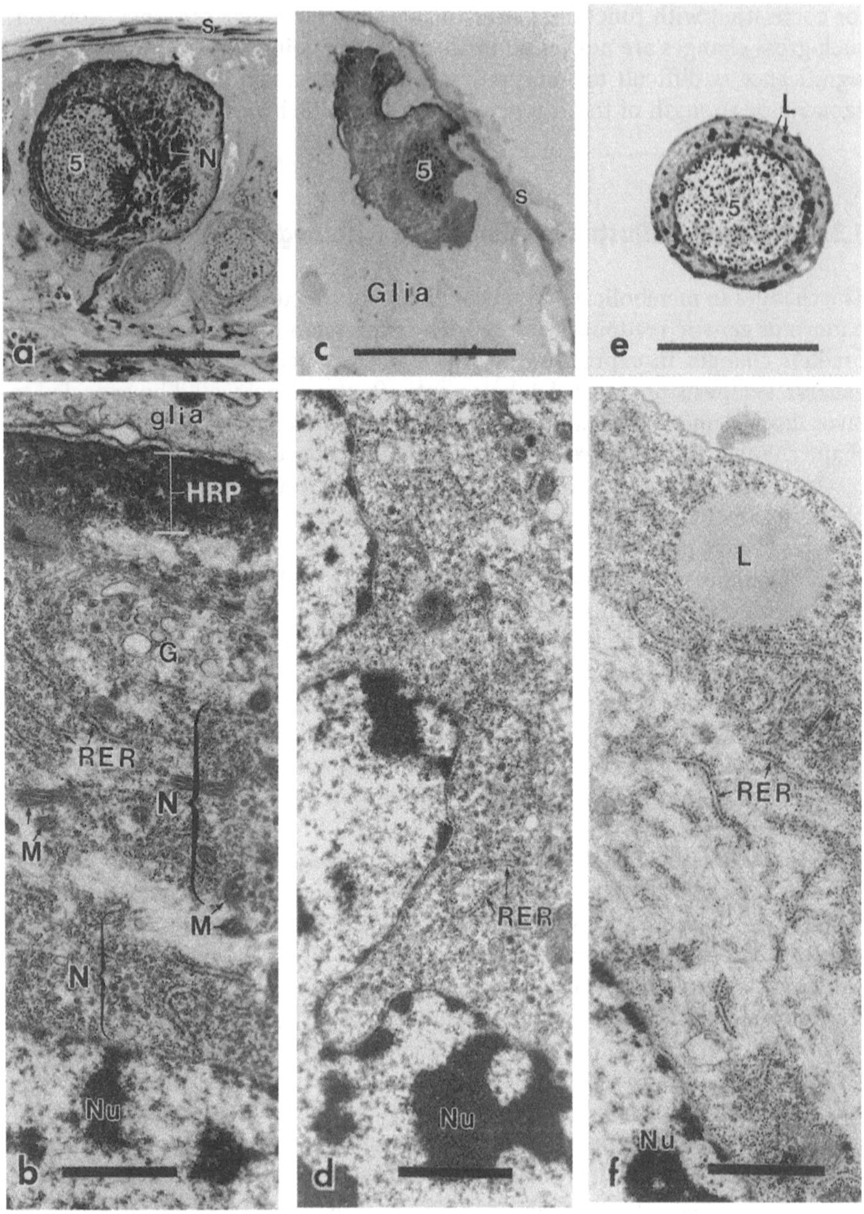

Fig. 3.10. Light and electron micrographs from sections of *Helisoma* buccal neuron 5. **a** Neuron 5 within an acutely dissected buccal ganglion. *N* Nissl bodies; *S* connective tissue sheath. **b** Perinuclear cytoplasm of neuron 5 from *a*. Note that the *HRP* was only visible in the periphery of the cytoplasm. *N* adjacent to the *brackets* denotes the approximate boundaries of dense accumulations of rough endoplasmic reticulum and associated polysomes that were correlated with Nissl bodies in light micrographs. *a* few mitochondria (*M*) were also visible. *RER*, rough endoplasmic reticulum, *Nu*, nucleus. **c** Neuron 5 five days after axotomy and organ culture.

3.3.5 Under What Conditions Is Regeneration Initiated?

Invertebrate neurons usually survive axotomy. However, initiation of regenerative growth is quite variable, reflecting intrinsic differences in the neurons, lesion site, age and physiological state, and trophic signals. Neurons consistently respond more dramatically if the injury is close rather than far from the soma. Perhaps electrical changes associated with injury sum with the altered retrograde message if the latter arrives soon enough. Distant injuries may also be less effective because the injury signal is decrementally transported or is contradicted by "intact" signals supplied by neural or nonneural cells through remaining contacts.

In the snail *Helisoma*, repair is not initiated following transection of the ipsilateral axon of buccal neuron B4, as long as the contralateral axon is intact (Kruk and Bulloch 1992; Fig. 3.11). In this case, the remaining segment of the injured axon is absorbed into the soma. In contrast, a more distal injury is followed by regeneration. In the light of the events outlined for retrograde signals generated by axotomy in *Aplysia* sensory neurons (Walters and Ambron 1995), the complete removal of an axon might remove the molecules that constitute an altered signal following axotomy.

The injury signal can be increased and the latent time decreased by crushing and then crushing again at a later time, an established conditioning-lesion procedure in vertebrate regeneration (rat sciatic or fish optic nerve: McQuarrie et al. 1977; McQuarrie and Grafstein 1982). The stimulation of axonal outgrowth by conditioning lesions is attributable to acceleration of the component of axonal transport that carries cytomatrix proteins (Jacob and McQuarrie 1993). In *Aplysia*, the response to cyclic adenosine monophosphate (cAMP) (which may be elevated by serotonin in response to injury) includes increases in fast axoplasmic transport as well as retrograde transport (Azhderian et al. 1994). A conditioning lesion is effective in stimulating growth in an *Aplysia* cerebral neuron that otherwise does not regenerate (Fredman and Nutz 1988).

3.3.6 Speculation on the Role of Diffusible Factors in Regeneration

Given that the metabolic machinery of the individual neuron is activated by injury and that the stage is set for regeneration, what role is played by diffusible growth promoting factors? Can differences between young and old animals in

---

(Fig. 3.10 continued:)
Dense staining Nissl bodies were not visible in the cytoplasm. Diffuse cytoplasmic basophilia was darker in neuron 5 five days after axotomy, but the basophilic staining was not as apparent in black and white prints as it was in sections. **d** Same as **c**, note *RER* and polysomes, although visible, were not as tightly arranged. **e** Isolated neuron 5 after 5 days in vitro. Dense-staining Nissl bodies such as those in **a** were not visible in the cytoplasm. *L* Lipid-like bodies. **f** Same as **e**. Note the lipid-like body *L* and loose arrangement of *RER* in the perinuclear cytoplasm. Magnification *bars* in **a, c, e** = 100 μm; **b, d, f** = 1 μm. (Berdan et al. 1990)

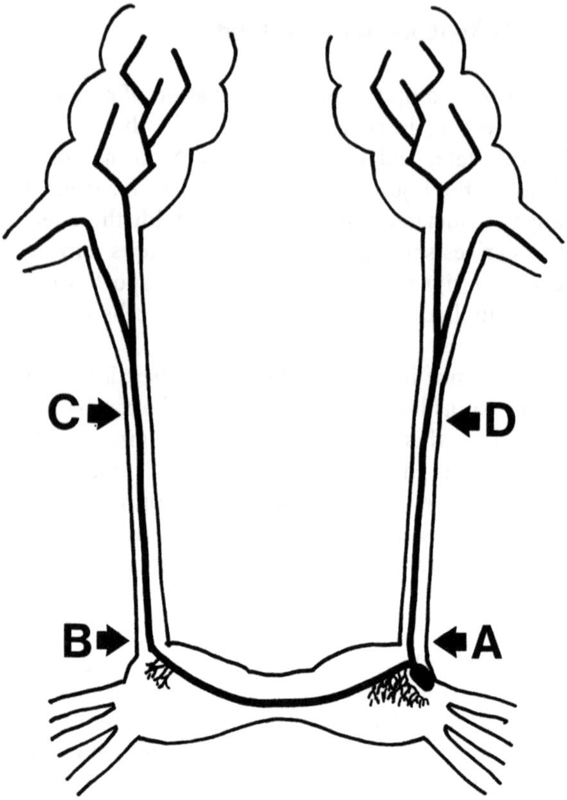

Fig. 3.11. Schematic diagram of *Helisoma* neuron B4 in the buccal ganglia, with its bilateral projections extending through the esophageal nerve trunk to the salivary glands. The axons were transected at the sites indicated (**A-D**). If an axon is transected at *B*, *C* or *D*, regeneration is initiated, but if it is only transected at site **C**, no regrowth is seen. (Data from Kruk and Bulloch 1992)

the rate and extent of repair reflect differences in growth factors? Young animals regenerate more easily and generally have higher levels of growth hormone than older animals. For instance, regeneration of crushed nerves in *Aplysia* is significantly faster and more effective in young animals (M. Scott and S. Fredman, pers. comm.).

Blood and nervous systems of many invertebrates contain insulin-like hormones (e.g., Sonetti and Bianchi 1993; Reinecke et al. 1993). The responses of cultured gastropod neurons to nerve growth factor suggest that additional factors in the blood may contribute importantly to regeneration in that group (Ridgway et al. 1991; Chap. 4). Studies of tentacle regeneration in the land snail *Achatina fulica* suggest that a blood-borne factor can speed regeneration and a different factor can slow it (Sidelnikov and Stepanov 1991, 1992; Moffett et al. 1995). The titer of such factors is elevated following limb autotomy in crustaceans, and observations on annelids (Chap. 2) suggest that they will be a fruitful source of information on such effects.

3.4 Survival of Anucleate Axon Segments

3.4.1 Mechanisms of Distal Segment Survival

Typically, the distal (anucleate) portion of a severed axon degenerates rapidly and the resulting cellular debris is removed by phagocytosis. This Wallerian degeneration is seen in mammalian neurons, crustacean sensory neurons, and insect motor neurons, to name a few. The pattern is not surprising, considering that distal segments lack the synthetic support provided by the nucleus. The remarkable thing is that in a number of animals the distal segments of some classes of neurons survive, may sprout, and may maintain connections to target cells (Bittner 1988).

When isolated axonal segments survive, it is usually with the help of other cells. Transfer of metabolites from glia is seen in intact systems (Lasek et al. 1974; Smith and Howes 1987), and large molecules can be exchanged in vesicles released by an intact cell and picked up by the axon segment (Viancour et al. 1988). A more surprising mode of survival is provided by incorporation of glial nuclei, as described below.

In vivo, *Aplysia* distal axon segments survive for weeks to months (Ross et al. 1994).. When this phenomenon is studied in vitro, survival time is shorter, but coculture allowing direct physical contact with homologous intact neurons approximately doubles survival time, from 7.6 to 15.3 days (Benbassat and Spira 1993, 1994). The axon segment does not synthesize protein, but relies on pools of protein and energy stores present prior to transection to support normal membrane properties and sprouting. The anucleate axonal segment forms electrical connections with intact homologous neurons or the proximal segment of the same cell, but not with distal segments of homologous neurons (Benbassat and Spira 1993).

The distal portions of crustacean motoneurons survive for many months (Hoy et al. 1967). Metabolic support of the distal segment by glia occurs (Viancour et al. 1988) and the satellite (glial) cells may invade, leaving their nuclei in the axonal cytoplasm (Atwood et al. 1989b; but see also Sheller et al. 1991).

3.4.2 Consequences of Distal Segment Survival

3.4.2.1 Target Support by Surviving Distal Segments

Long-term survival of distal segments of severed axons potentially offers several advantages. First, the target may not undergo denervation-induced alterations. Despite the fact that the distal segment of a motor neuron would not exhibit the pattern of electrical activity normally generated by the CNS, there is ample evidence for the extra-synaptic release of transmitter (Grossfeld et al. 1994). A surviving distal motor axon segment may continue to occupy synaptic terminal sites and forestall reactive synaptogenesis.

The distal segment of severed crustacean motoneurons must be fairly long in order to survive, because distal portions of axons severed close to the muscle cannot survive and spare target muscles from degeneration (Bouton and Bittner 1981). The distal portion remains competent to conduct action potentials and initiate muscle contractions for at least a year (Atwood et al. 1989a). The mechanism that controls the time course of vesicle release is altered in the terminals of these decentralized axons, perhaps because the glial synthesis does not include appropriate proteins to support this neuronal function (Parnas et al. 1991).

In the response of the whole crayfish connective (including axonal and nonneuronal cells) to crush, changes in the protein synthesis pattern within the first few hours include expression of a 70 kDa protein, related to mammalian stress protein, that is maintained for at least a month (Xue and Grossfeld 1993). These changes may reflect the responses of nonneuronal cells attracted to the site of the injury and subsequently associated with the maintenance of motoneuronal segments severed from their somata.

3.4.2.2 Pathway Preservation

Another advantage of distal segment survival is that regenerating axons can grow along distal segments to the target, increasing the probability that the original innervation pattern is recreated. Distal axon survival is seen in at least some neurons in flatworms, crustaceans, insects, annelids, gastropods, and vertebrates (Koopowitz and Holman 1988; Bittner 1988). The survival and fusion of neural elements in cnidarian regeneration is a possibility that has not been carefully examined. Sprouting by the distal stump is reported for flatworms (Koopowitz and Holman 1988) and annelids (Lyckman et al. 1992). Temperature has a large effect on the survival of the goldfish Mauthner axon, as is generally true for vertebrate and leech distal axonal processes (Cancalon 1985; Carbonetto and Muller 1982), whereas survival of the crayfish giant axon is independent of temperature (Blundon et al. 1990).

3.4.2.3 Proximal and Distal Segment Fusion

The overwhelming advantage conferred by distal segment survival is quick return of function when an electrical link is reestablished between the severed parts of the neuron, restoring communication with the target. The type of contact that can form between the axon segments differs in different animals and a range of responses is displayed by different neurons in the same animal (Fig. 2.11; Bittner 1991). Cytoplasmic continuity produced by fusion of the proximal and distal segment membranes around the core of axoplasm is probably the ideal in repair, because it is sufficient to rapidly restore both the neuron's integrity and its metabolic communication with postsynaptic elements.

In the flatworm *Notoplana acticola*, cut ends of the major longitudinal nerve appear to rejoin in a rapid (36–48 h) recovery of function after transection (Koopowitz et al. 1975, 1976). Brain neurons filled with Lucifer yellow dye extend filopodia toward the distal stump and these contact a number of axons, but have never been observed to establish cytoplasmic continuity with more than one. Once cytoplasmic continuity is established, the diameter of the link increases and other filopodia disappear. The question of specificity of reconnection has not been directly addressed, but individual axons that fail to contact a target within the normal period (14 h) exhibit extensive filopodial projections in what might be interpreted as a pattern of searching for a particular target (Koopowitz and Holman 1988).

Axonal fusion is established on morphological grounds for leech mechanoreceptor neurons (DeRiemer et al. 1983). This electrical continuity will restore neuronal function but may not allow the metabolic transfer necessary to support the distal segment. For whatever reason, this temporary measure is often followed by continued growth of processes from the soma-attached part of the cell to the true target, following the guidance provided by the "rescued" distal segment. Once regenerating processes synapse with the target, the distal part of the old axon loses the electrical synapses it received from the proximal segment and degenerates. Similarly, in crustaceans, formation of electrical contacts restores specific reinnervation immediately (Hoy et al. 1967); this may be followed by the regrowth of processes along the distal segment to the target. Such a sequence allows a rapid return of function, as well as accurate recreation of the original projection pattern and connectivity.

In leeches, S cells connect with S cells of adjacent ganglia by electrical synapses in the center of the connective, and S cells can also form a functional connection between the proximal and distal portions of an individual axon (Carbonetto and Muller 1977; Muller and Carbonetto 1979). After connective crush close to a ganglion, the existing electrical synapse continues to function and repair involves formation of a basket-like series of contacts between the cell-attached part of the neuron and its anucleate segment in the injured region of the connective. Over time, the repair site grows closer to the normal site of the electrical synapse between the two cells.

The capacity for axon segment reconnection in leeches has been explored with photoablation (Camhi and Macagno 1991). This method of breaking an axon has the advantage of not disrupting surrounding tissue. For instance, when the continuity of a Retzius cell axon is destroyed by irradiating a small portion of the dye-filled axon, the rest of the neural and sheath elements are not disrupted. Under these circumstances, 70% of the proximal and distal axon segments become electrically connected. Formation of the junction was tested by comparing the spread of Lucifer yellow between proximal and distal segments within 3 days of the lesion (Fig. 3.12). These results were contrasted with the failure of the much larger horseradish peroxidase molecules to spread beyond the injected side of the junction, suggesting that cytoplasmic fusion had not occurred. The success of point ablation in this system contrasts with the disorderly sprouting and lack of reconnection following a nerve crush.

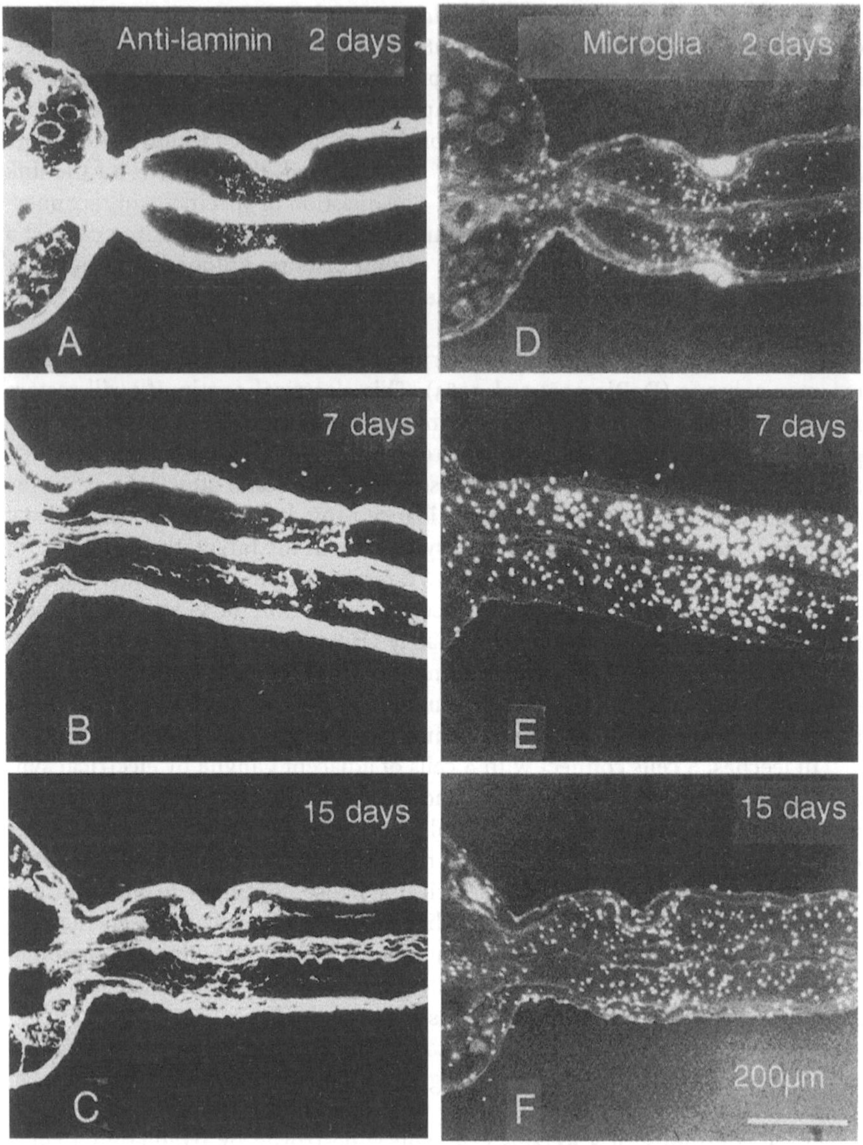

Fig. 3.15. Longitudinal sections of leech CNS stained for laminin (*left*) and microglia (*right*). The connectives had been crushed 2 days, 7 days and 15 days previously. Part of the ganglion is seen to the left of each photo. **A-C** Laminin was labeled by mAb 206 and rhodamine conjugated antimouse IgG (*antilaminin*) and **D-F** nuclei were labeled with Hoechst stain (*microglia*). Two days after a crush, laminin appeared within the connectives at the crush site (**A**) and microglial cells accumulated both outside and within the connectives (**D**). Seven days after a crush, laminin aggregated into clumps and streaks at the site of injury (**B**) and microglial cells remained within the connectives at the crush site (**E**). Laminin stain was fibrillar 15 days after a crush (**C**) and microglia were more uniformly distributed along the connectives (**F**) (Masuda-Nakagawa et al. 1990)

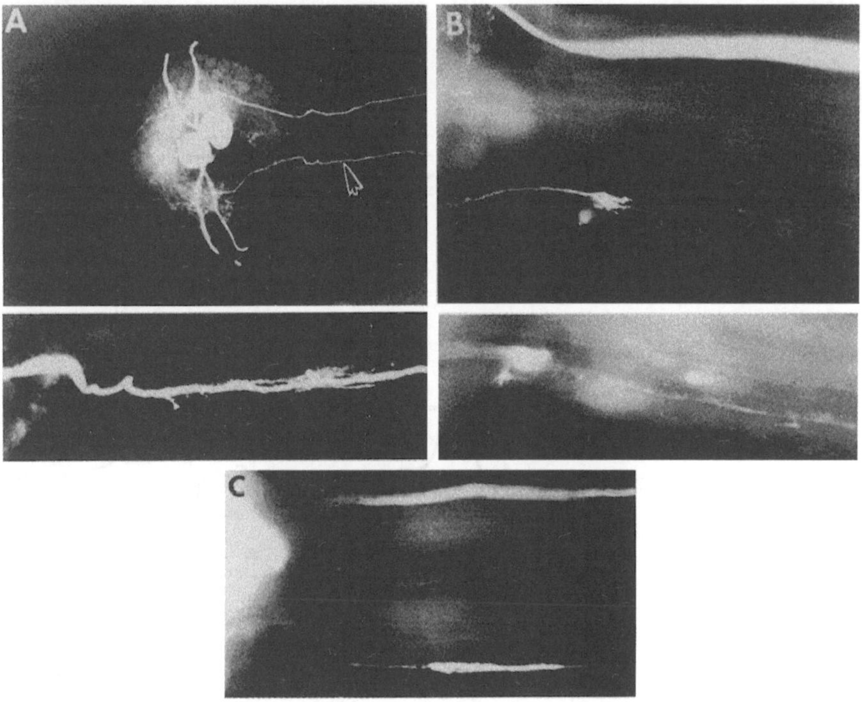

Fig. 3.12. Regeneration in vivo of photoablated Retzius axons. **A** (top panel) Four days after photoablating the anterior axon of one Retzius cell, both this cell and its contralateral homologue have been filled with Lucifer yellow. Anterior is to the *right* on this and all other photos in this figure. All four axons of each of the cells (one anterior, one posterior, and two peripheral) are filled. The region of the prior photoablation is indicated by an *arrow*. **A** (*bottom panel*) Enlargement of the photoablated region. Only in this region are multiple sprouts seen; a single large diameter axon is also filled. **B** (*top panel*) A different regenerating Retzius axon. A control axon can be seen at the *top* of the photo, with the photoablated axon *below*. Several sprouts are seen. **B** (*bottom panel*) In a different plane of focus one of these sprouts is seen to contain a Lucifer yellow-filled axon. **C** A third example showing yet a different shape. (Camhi and Macagno 1991) Scale *bar* = approx. 200 μm in **A**, 50 μm in **B** and **C**, 20 μm in enlargements of **A** and **B**

In the earthworm *Eisenia foetida*, neural regeneration is more robust than in *Lumbricus terrestris* and the anatomical repair more closely approaches the control condition. Transection of the nerve cord in *Eisenia* is followed by recovery of appropriate withdrawal responses within 1–2 days; if a segment of nerve cord is grafted from another animal, the reconnection requires several days, with repair seen in the lateral giants before the medial giant (Drewes et al. 1988). These experiments monitered recovery in intact animals, utilizing action potential waveform and the pattern of recruitment to suggest that repair involves specific reconnection of proximal and distal segments.

Because earthworm medial (unpaired) giant axons (MG) and lateral (right and left) giant axons (LG) are multicellular structures, distal segments isolated

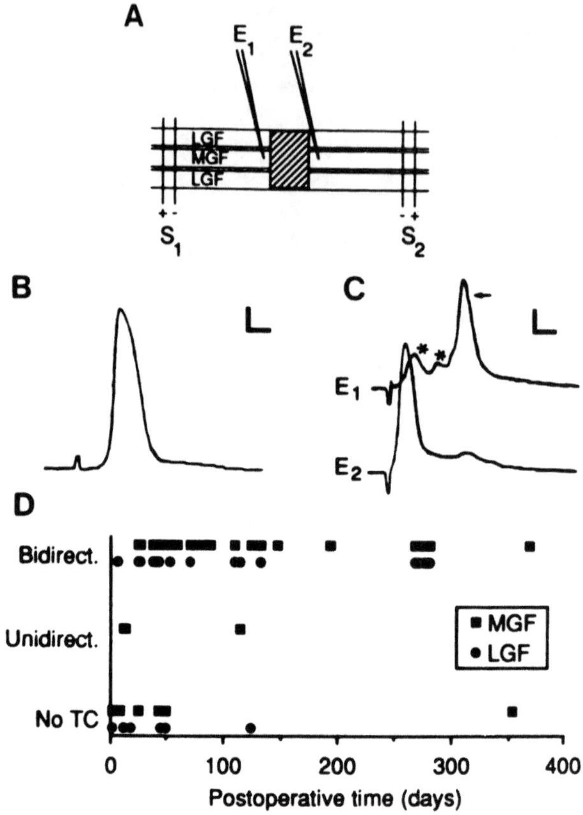

Fig. 3.13. Control and through-conducted action potentials. **A** Configuration of recording and stimulating electrodes. *E1, E2* Intracellular microelectrodes for recording or stimulating. *S1, S2* Extracellular stimulating electrodes. **B** Action potential in control medial giant axon. Vertical calibration, 13 mV; horizontal calibration, 0.4 ms. **C** Responses 41 days posttransection. Lower trace (*E2*), action potential following stimulation caudal to the lesion site (*S2*). Top trace (E_1), through-conducted action potential in the medial giant axon (*arrow*) recorded rostral to the lesion site. *Asterisks* denote prepotentials. Vertical calibration, 20 mV; horizontal calibration, 1.2 ms. **D** Incidence of appropriate, through-conducted action potentials (bidirectional, unidirectional) for lesioned medial and lateral giant fibers versus postoperative time (days). *No TC* no through-conduction. (Lyckman and Bittner 1992)

from their own cell body are potentially the recipients of metabolic and synthetic support through their continuity with segments that have somata (Lyckman et al. 1992). Specificity of reconnection of giant fibers has been convincingly documented in *Lumbricus* by intracellular recordings of conduction across the lesion site (Fig. 3.13 Lyckman and Bittner 1992). The conduction initially involves a significant delay and a low safety factor; these probably reflect the fact that the connection is formed by one or several small neurites. Selectivity is apparent from the first evidence of electrical conduction, although anatomical evidence of dye coupling to nongiant neurons is occasionally seen (Lyckman et al. 1992).

Although there are no reported instances of gastropod axons fusing in vivo, and the alternative response has been documented in some cases (Ross et al. 1994), fusion occurs in cultured neurons (Benbassat and Spira 1993, 1994) and this is therefore a possible mechanism in vivo, if the anatomical situation is favorable.

3.4.2.4 Experimental Approaches to Axonal Fusion

One experimental approach to the repair of axons involves tightly apposing the cut ends in the presence of polyethylene glycol (Krause and Bittner 1990; Krause et al. 1991). Morphological fusion of the medial giant axon of the earthworm is achieved within minutes by this method, which may be attributed to the dehydrating and membrane-altering actions of polyethylene glycol (Parente and Lentz 1986). Lucifer yellow injected into one medial giant axon segment spreads across the lesion site to heavily label the opposite segment (Krause and Bittner 1990). Ultrastructural examination of the fusion site shows disorganized axoplasmic constituents, but a continuous cytoplasm across the site of fusion.

Other methods for promoting rapid reconnection of severed medial giant axons include laser pulses (Yogev et al. 1991) and the application of an electrical field (Todorov et al. 1992). Focused laser pulses achieve reconnection with a 40% success rate. The electrical field technique, which has been used to produce heterokaryons (Zimmerman 1986) involves placing anode and cathode in the axon segments across the break and passing electrical pulses of 80–200 V for 10–100 usec. The success of reconnection is confirmed by Lucifer yellow diffusion across the lesion site, and light and election microscopy (Fig. 3.14).

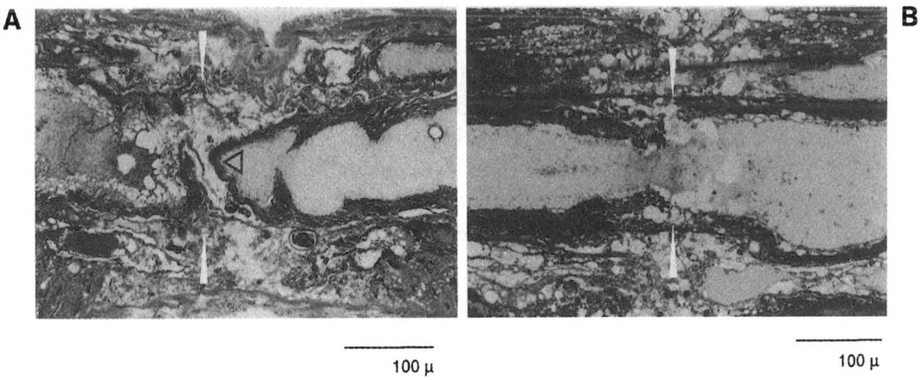

Fig. 3.14. Electric field-induced reconnection of severed earthworm medial giant axons. **A** Control (untreated) axon after the nerve cord was crushed with a razor blade. Glial cell sheaths covering the rostral and caudal axons (*open black arrowhead*) demarcate the severed segment of the MGA (*solid white arrowheads*). No axoplasmic continuity exists between rostral and caudal axons. **B** Axon reconnection at the region of the *white arrows*. After 3–5 electrical pulses of 80–200 V and 10–100 μs, the axoplasm of the rostral and caudal segments of the medial giant axon appear to be in continuity. Ultrastructural examination revealed no membrane partition. (Todorov et al. 1992)

3.5 Responses of Nonneural Cells to Injury

Crushing a nerve with forceps severs axons and also may kill some nonneural cells in the sheath, but sheath connective tissue is often intact enough to serve as a pathway for growing axons to follow. The role of sheath elements in promoting neural outgrowth in culture reflects its chemical composition, which is presumably attributable to synthesis in nonneural cells (Miller and Hadley 1991). Often nonneural cells move into the site of damage before any outgrowth by regenerating neurites, creating an environment that is crucial for axonal regeneration. Repair cells can originate from glia in the neural sheath or the blood.

Leech microglia begin to migrate into the crush site from other sites in the sheath within minutes (McGlade-McCulloch et al. 1989). They are present on the surface of the crushed connective whether or not the nervous system is isolated, so bloodborne factors are not required for the attraction (Morgese et al. 1983; McGlade-McCulloch et al. 1989; McGlade-McCulloch and Muller 1989). Changes in the characteristics and distribution of extracellular matrix molecules are correlated with the microglial responses (Fig. 3.15; Masuda-Nakagawa and Nicholls 1991; Masuda-Nakagawa et al. 1994). The number of microglia in the crush region and microfibrillar staining of laminin increase in parallel during the weeks following the lesion (Masuda-Nakagawa et al. 1990). The laminin is located not only among the elongating axons but also ahead of the leading axons, which suggests that it is not released from growing axons (Masuda-Nakagawa and Nicholls 1991). The laminin in contact with the growing neurites is punctate, which also associates it with the spotty distribution of the microglia, in contrast to the situation in the intact connectives, where laminin is restricted to the basement membrane of the endothelial layer and capsule matrix, and is not in contact with axons (Chiquet et al. 1988; Masuda-Nakagawa et al. 1990). The laminin is clearly associated with the microglia, but it is not the only adhesion factor that controls neurite outgrowth behavior (Masuda-Nakagawa and Wiedemann 1992; Masuda-Nakagawa et al. 1994).

After a cockroach peripheral nerve is cut, hemocytes accumulate at the end and penetrate the nerve, where they exhibit phagocytic activity (Fig. 3.16; Blanco 1990; Blanco and Lane 1990). The role of glia in creating an environment for regenerating axons has been documented at the ultrastructural level. A similar pattern of tunnels through which growing neurites project is seen after cerebral commissure transection in snails (Moffett, unpub.). Depending on whether or not glia are present, the growth of axons from explanted cockroach ganglia follows different patterns. Neurites lacking glia form a diffuse network quite different from the nerve-like processes of axons bundled within a glial-like sleeve (Howes et al. 1989).

Damage to sheath cells and neurons in insects has been experimentally separated by chemical lesions that leave the axons intact. The DNA intercalating drug ethidium bromide has been applied to selected regions of connectives to kill the glia (Smith and Howes 1987). In many insects, the blood K^+ concentra-

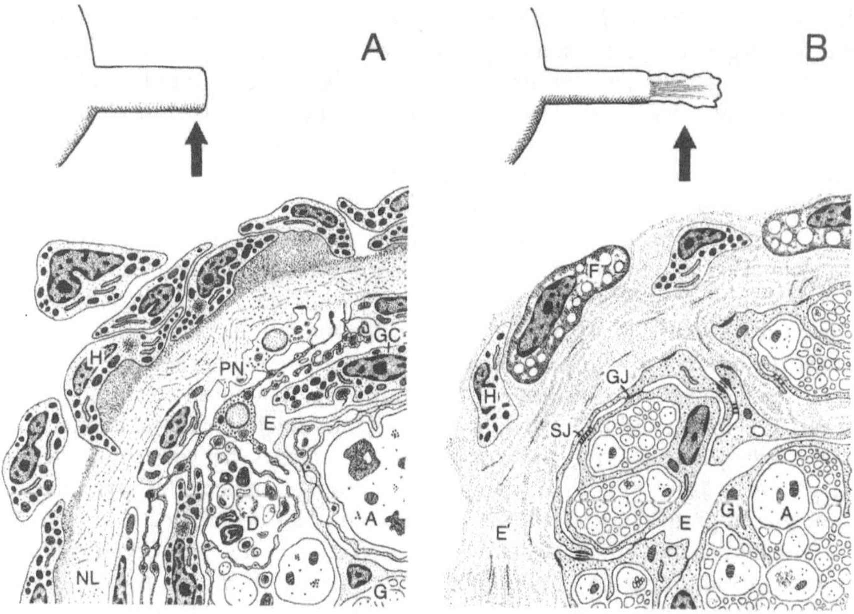

Fig. 3.16. Diagrams summarizing the appearance of transverse sections of the proximal stump of the cut metathoracic nerve of a cockroach. **A** Four days after nerve transection. Hemocytes (*H*) cluster around the outside of the neural lamella (*NL*). Some perineural cells (*PN*) show signs of degeneration. Granule-containing cells (*GC*) are present under the perineurium. Many small axons have degenerated (*D*) but large axons (*A*) and glia (*G*) are still present. Clumps of extracellular material (*E*) are present in the enlarged extracellular spaces. **B** Four weeks after nerve transection the outgrowth from the nerve stump is composed of fascicles of axons (*A*) surrounded by glia (*G*) and embedded in extracellular material (*E*) containing collagen fibrils. Septate junctions (*SJ*) and gap junctions (*GJ*) are made between the glial processes, and large deposits of extracellular material are present between those processes (*E*). A few hemocytes (*H*) and fat cells (*F*) are present outside the outgrowth. (Blanco 1990)

tion is so high that a tight blood-brain barrier must be created by the sheath cells following injury before action potential propagation can resume. Initial invasion of the damaged region is by hemocytes, which accumulate on the surface of connectives within 48 h after glial disruption (Treherne et al. 1987, 1988). The hemocytes penetrate the sheath and are transformed into granule-containing cells; they subsequently disappear as a class and are likely to be transformed into sheath components (Smith et al. 1991). Their contribution to sheath repair involves both structural and inductive roles, since they recruit endogenous glial proliferation following their invasion (Smith et al. 1990). Once structural aspects of sheath repair are complete, the potassium permeability characteristic of the normal sheath is restored and action potential amplitude recovers. Hemocytes support the sprouting of adult neurons in culture and may similarly support in situ regeneration (Howes et al. 1991; Chap. 4).

In gastropods, both hemocytes and glia respond to nerve crush. At the ultrastructural level, two types of glia are distinguishable in association with the buccal ganglia of *Helisoma*: 90% are cells with electron-dense cytoplasm and few organelles; the remaining glia have more electron-lucent cytoplasm, numerous organelles, and an irregular outline that suggests motility or phagocytic function (Berdan et al. 1989). The latter type, which is similar to hemocytes which attach to the ganglion surface or individually cultured neuron cell bodies and neurites (Berdan et al. 1990), may represent a cell type that moves readily between the blood and the nerve sheath.

In the snail *Melampus*, cells accumulate at the margin of the crush site along a nerve or connective (Moffett and Bale 1993; Fig. 3.17). In an intact neural sheath examined two days after injection of a phagocytic marker, there are rarely cells attached that have accumulated the marker, but such cells accumulate at the lesion site within 4–8 h and continue to be found there for at least 7 days. Some cells at the crush site exhibit mitosis, as evidenced by accumulation of BRDU (Moffett and Bale 1993), whereas glia in intact nerves and connectives have a low mitotic activity. The scenario that emerges is that cells committed to phagocytosis prior to the nerve injury (i.e., probably hemocytes) accumulate, some glia migrate along the sheath and some mitosis occurs there.

The initial stage of tentacle and eye regeneration in the snail *Cryptomphalus aspersa* includes aggregation of amoebocytes on the stumps to form a bridge between tissues that subsequently grow together (Fig. 3.18; Flores et al. 1992).

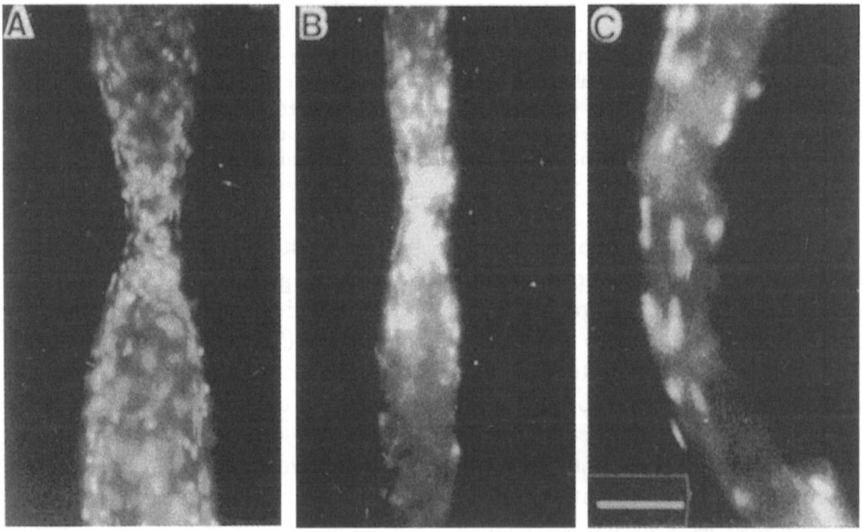

Fig. 3.17. Responses to nerve crush in *Melampus*. **A** Sheath cells (microglia), revealed by Hoechst nuclear stain, 3 days following the crush. **B** Phagocytic cells, characterized by accumulation of a fluorescent marker (Sigma PKH26) injected into the blood 2 days before the crush have accumulated along the crushed region within 12 h. **C** Cell division is evident in the incorporation of bromodeoxyurudine; crush was 9 days before sacrifice. Scale *bar* =40 µm. (Moffett and Bale, unpubl.)

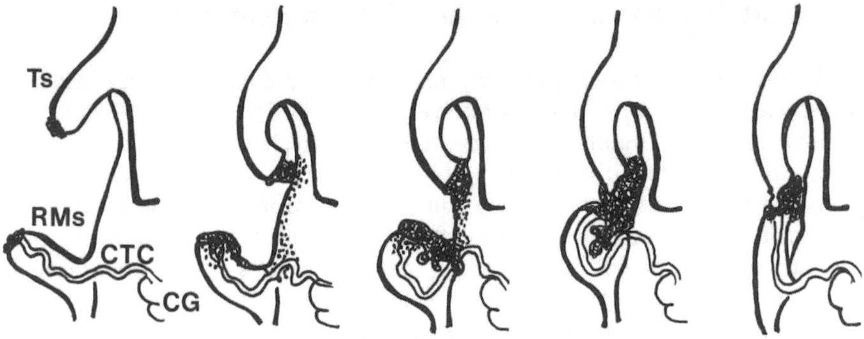

Fig. 3.18. Schematic representation of the blastemal reconnection process in tentacle regeneration. *Black areas* represent amoebocyte aggregation; *dotted areas*, amoebocytic infiltration of the coelomic sheath. *CG* Cerebral ganglion; *CTC* cerebrotentacular connective; *Ts* tentacle stump; *RMs* retractor muscle stump. (Redrawn from Flores et al. 1992)

Settling and transformation of hemocytes is thought to provide the replacement cells for ganglion regeneration in tunicates (Bollner 1989; Bollner et al. 1993). An attractant is apparently involved in the movement of hemocytes to a site of neural injury. Settling and conversion of the hemoctyes during regeneration may be initiated by an immune response to injury, which is well established for both vertebrates and invertebrates (Stefano 1992; Millar and Ratcliffe 1994). There is evidence in human beings that higher brain functions may coordinate such responses, as illustrated by a study of soldiers training in parachute jumps, who exhibit an increase in nerve growth factor in the bloodstream prior to a jump, even before increases in cortisol and adrenocorticotrophic hormone are seen (Aloe et al. 1994). In the invertebrates, the presence of ACTH in neurons and hemocytes and the related melanocyte stimulating hormone, MSH, causes hemocytes that are exhibiting ameboid activity to become quiescent and inactive (Smith et al. 1991).

After a crayfish abdominal connective is crushed, the site is invaded by cells of unknown origin that replace the volume of the severed axons; adjacent regions with intact axons also show accumulations of cells, but more distant regions are unaffected (Grossfeld and Hansen 1987). When the nerves innervating crayfish limbs are severed, the soma-less central axon stumps of sensory cells (with somata in the periphery) degenerate over approximately 3 weeks, in contrast to the relatively long-term survival of the motor axons (Bittner 1988; Govind et al. 1992). In nerve cross sections, the axons are apparently invaginated by glia and an occasional nucleus resembling glial nuclei is seen in otherwise intact axons. Support for the long-term survival of rock lobster motoneurons by glia is correlated with the incorporation of glial nuclei (Atwood et al. 1989b), so the glial nuclei in sensory axons may represent a similar, if less successful, response.

3.6 What Activates Responses in Uninjured Neurons?

We have seen that neurons that are injured receive a number of cues that may shift them into the repair mode. However, many physically intact neurons also respond to neural trauma and tissue damage by sprouting. Such changes are most easily detected in nervous systems in which the detailed anatomy of identified neurons is known. Molluscan neurons sprout following a variety of indirect insults. Injury to the neural sheath can lead to sprouting in the presumed absence of neural injury (Altrup 1982). Injury to a central tract through which the *Lymnaea stagnalis* interneuron RPeD1 does not project elicits sprouting and novel projections from that neuron (Allison and Benjamin 1985). Three possible contributions to the sprouting response were suggested: injury substances, loss of presynaptic inputs, and removal of synaptic competition (Benjamin and Allison 1985, 1987).

Axotomy is one of the most reliable triggers for sprouting for many neurons, but injury of nearby neurons can elicit sprouting of undamaged neurons, as illustrated by the insect neuron studied by Pitman and Rand (1982). Photoreceptor ablation results in reactive synaptogenesis which converts contacts that are dyadic to single, monadic synapses in the fly optic lobe (Brandstätter et al. 1992).

Target removal or loss of synaptic inputs can be potent stimuli for sprouting of uninjured neurons. In the leech, some neurons respond to indirect effects of injury. For instance, removal of one of the electrically coupled S cells does not trigger sprouting until the remaining S cell is subjected to axotomy (Muller and Scott 1980), but the N cell sprouts in response to peripheral or central target loss more readily than to axotomy (Bannatyne et al. 1989; Muller and Gu 1991). Neurons in the snail *Achatina fulica* sprout following the partial depletion of serotonin with 5,7-dihydroxytryptamine and p-chlorophenylalanine (Baker et al. 1993). Serotonin halts the extension of growth cones of some *Helisoma* neurons (Haydon et al. 1984). Similarly, when serotonin is absent during development, neural morphology is altered in *Helisoma* and *Drosophila* (Budnick et al. 1989; Goldberg and Kater 1989).

Blood loss and exposure to hypertonic environments can cause temporary changes in neuron morphology (sprouting) in some *Helisoma* neurons (Bulloch 1984; Maetzold and Bulloch 1986). Both of these stresses involve elevation of blood acidic amino acids, such as glutamate and aspartate (Jones et al. 1987), which may in turn lead to the release of trophic factors (Bulloch 1989). (Neurotrophic factors are discussed in Chapter 4). Deafferentation has been little studied in gastropods, and details of changes in central circuitry in the period preceding reinnervation are unknown.

In the leech, a large glial cell normally ensheaths all axons in a lateral connective. Ablation of this glial cell causes uninjured neurons to sprout; axotomy and glial cell destruction are additive, resulting in more vigorous sprouting (Elliot and Muller 1983). Microglia replace the large glial cell, ensheathing the new growth (Elliot and Muller 1982).

A lesion at one level of the leech nervous system can not only alter the balance of excitation and inhibition between regenerating sensory to motor connections, but also has similar permanent effects on homologous connections between uninjured neurons in distant ganglia (Baylor and Nicholls 1971; Jansen and Nicholls 1972). These distant alterations may result from the removal of inputs from through-conducting axons. Destruction of a single motoneuron allows homologous neurons in adjacent ganglia to innervate the denervated target muscle (Bowling et al. 1978; Blackshaw et al. 1982).

Denervation is also an effective stimulus for the formation of new central contacts. Leech N (nociceptive) cells project to innervate the cells of other ganglia, and the homologs of their normal target cells in their own ganglia are innervated by N cells from adjacent ganglia. However, if the inputs of N cells from adjacent ganglia are removed, targets in the same ganglion can be innervated (Gu and Muller 1990; Muller and Gu 1991). In some instances, a combination of denervation and neuron injury is necessary to elicit sprouting (Muller et al. 1987).

3.7 Conclusions

The earliest responses to injury are the depolarization and electrical discharges that reflect membrane disruption. These may send a rapid signal to the soma and cause release of neuromodulators. At the site of injury, ion entry, particularly Ca^{2+}, can lead to many changes, including activation of enzymes that generate retrograde signals or alter the cytoskeleton so that membrane sealing can occur.

Ongoing processes such as axoplasmic flow can support growth-like responses in the short term, but the activation of altered synthesis is necessary to support significant outgrowth or the formation of new connections. Whether or not a neuron can regenerate is dependent upon a permissive genetic disposition and receipt of a sufficient signal in the soma. The initiation of a regeneration profile in the soma is only the first requirement for successful regeneration. Extrinsic signals discussed in the next two chapters support the process of regeneration. At each step of the way, the sufficiency of environmental cues and the innate capacity of the neuron to respond appropriately determine the success or failure of the venture.

Chapter 4

Pathfinding by the Growth Cone

4.1 Introduction

The growth cone is a specialization of developing and regenerating neurons which samples the environment and is attracted or repelled by the cues it detects. Different growth cones express different surface receptors or exhibit different second messenger responses to a given cue. The receptors expressed on the growth cone membrane change as the result of the contacts it makes. In a cell with several growth cones, each can assume a distinctive set of features that propel it along different anatomical pathways to different destinations.

Because it is the growth cone that reads the treasure map, written in molecular cues, and follows that map to the life-sustaining targets, intensive research has focused on growth cone behavior. Many excellent sources of information on the growth cone exist, including Kater and Mills (1991), Letourneau et al. (1991), Goldberg and Burmeister (1989), and Sabry et al. (1991).

4.2 Growth Cone Morphology

A growth cone is a well-defined enlargement at the end of a growing axon or dendrite. It contains cytoplasm and organelles surrounded by thin expanses of lamellipodia, tipped with a fringe of filopodia. Cultured mammalian, leech, or gastropod neurons are very similar; insect growth cones may have additional lateral projections (O'Connor et al. 1990; Sabry et al. 1991).

The organelle-rich central domain of the growth cone contains mitochondria, dense core vesicles, and microtubules that extend from the axon (Fig. 4.1). The microtubules are in a dynamic state: lengthening by addition of tubulin to the lamellipodial ends or disassembling. The microtubules of insect pioneer neurons, visualized by fluorescein-labelled tubulin, are not always straight; curves and loops are seen, particularly at branch points (Sabry et al. 1991).

The thin, peripheral domain of the lamellipodia and filopodia is loosely stabilized by a meshwork of actin microfilaments (Fig. 4.1). Bundles of microfilaments extend into most filopodia. Labelled actin monomers add onto the membrane-associated ends of the filaments; over time, labeled subunits move rearward toward the core (Okabe and Hirokawa 1991). The filopodia act as the sensors for environmental cues; any of the insect growth cone filopodia may

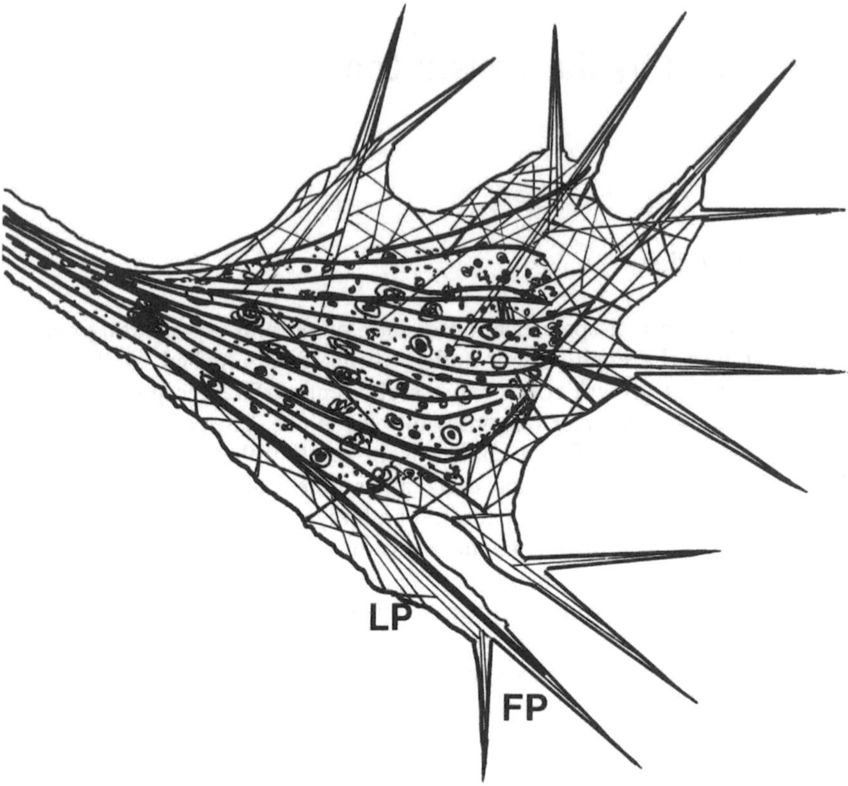

Fig. 4.1. The growth cone central domain is restricted by a microfilament network in the peripheral domain. Actin microfilaments extend through the lamellopodia (*LP*) into the filopodia (*FP*); microtubules run parallel in the axon and spread out in the central domain and sometimes extend into filopodia

become the leading projection if it encounters attractive cues (O'Connor et al. 1990; Sabry et al. 1991; Fig. 4.2).

Minutes after axon transection, filopodia begin to extend from cultured *Aplysia* neurons; approximately half of the filopodial growth can be blocked by blockers of actin polymerization and the other half by blockers of microtubule assembly (Goldberg and Burmeister 1992). Microtubules extend parallel to the microfilaments in some filopodia in cultured vertebrate neurons and rapidly invade advancing projections in insect pioneer neurons (Gordon-Weeks 1991; Sabry et al. 1991).

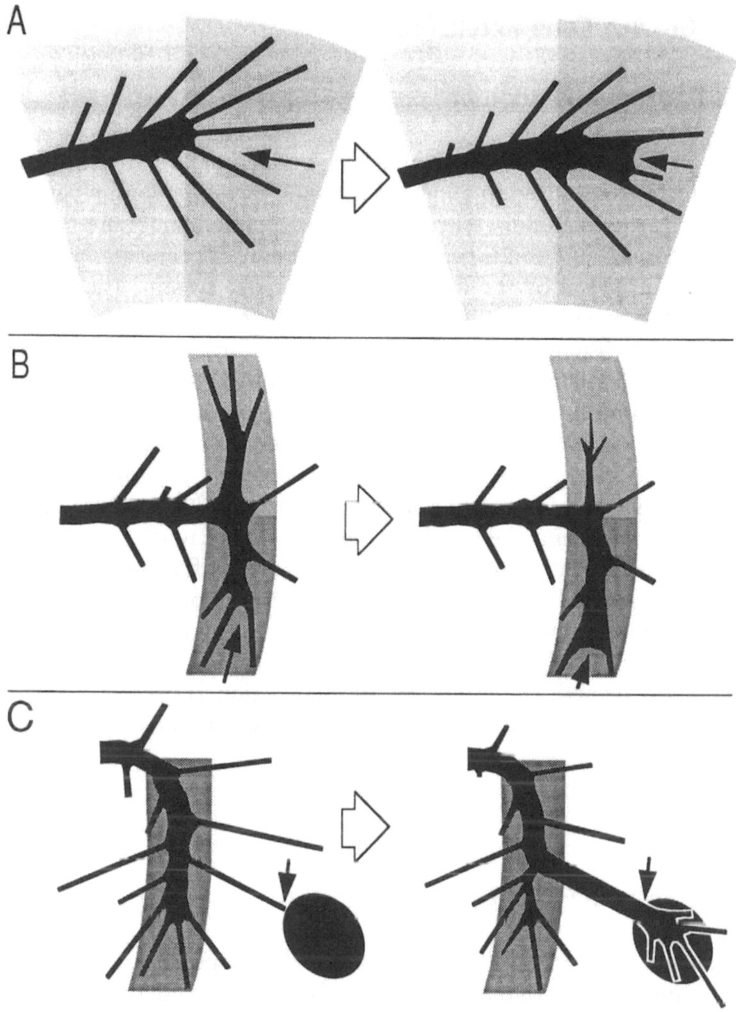

4.2. Diagrammatic representation of pioneer growth cone migration and steering events on in situ substrates. **A** Within the femur, the growth cones migrate between the basal lamina and the surfaces of epithelial cells. This substrate appears to be relatively homogeneous with a proximal gradient or a weak step increase in affinity. Growth cones advance by filling in flat veils between filopodia (veil extension) and do not advance along any particular filopodium for more than a few micrometers. Steering occurs by shallow angle turns along more proximally oriented filopodia, veils, or branches. **B** At the trochanter-coxa boundary, growth cones encounter an orthogonal band of segment-boundary cells which represents a sharp discontinuity in substrate adhesivity. This adhesiveness may increase ventrally. Growth cones usually extend major branches in both directions, with the ventral branch eventually being followed and the dorsal branch withdrawn. On the segment-boundary cells, the growth cone branches advance by veil extension. **C** At three locations in the limb, growth cones migrating on intrasegmental epithelial cells, or on the segment boundary cells, encounter preaxonogenesis (guidepost) neurons. Single filopodial contacts with any of these cells can abruptly reorient the growth cone. The steering event occurs by an increase in the diameter of the filopodium until it becomes a nascent axon (filopodial dilation). (O'Connor et al. 1990)

4.3 Growth Cone Extension

4.3.1 Membrane Addition

The growth cone grows as membrane is added at the extremity and the cytoskeleton is solubilized to allow cytoplasmic flow. Membrane patches are introduced by the fusion of vesicles with growth cone membrane. The rate and site of membrane addition is probably regulated by appropriate "addressing" of the vesicles delivered by the cytoskeletal transport system. The fusion process is thought to resemble secretory/synaptic vesicle release without the subsequent endocytosis (Pfenninger et al. 1990). The similarity of the two fusion processes is evident in a temperature sensitive mutant of *Drosophila*, which exhibits dysfunction in both synaptic transmission and growth cone elongation at temperatures above 30°C (Wu 1988). The immediate outgrowth in culture exhibited by peptidergic crustacean neurons may result from utilization of the abundant secretory granules in these cells, uncoupled from the usual membrane reuptake processes (Graf and Cooke 1990; Grau and Cooke 1992).

Goldberg et al. (1991) observed the sequence of movement of material from the central domain into the thinner membrane-bound lamella. Micropipette application of cyclic AMP or forskolin to one side of a growth cone of cultured *Xenopus* neurons produces a gradient within the growth cone and draws the advancing growth cone into the injected region (Lohof et al. 1992). In insect neurons, the extension is accompanied by microtubule projections (Sabry et al. 1991).

4.3.2 Reshaping the Cytoskeleton

The actin network in the lamellopodia and filopodia must be destabilized in order for elements of the cytoplasm-rich domain to advance into the peripheral region. A cytoplasmic tyrosine kinase characteristic of *Drosophila* CNS axons during embryogenesis may promote extension of growth cones (Gertler et al. 1989). Advancement of *Aplysia* growth cones and organelle redistribution has been studied with video-enhanced contrast-differential interference contrast microscopy developed by Allen et al. (1981). The lamellar peripheral zone of the growth cone restricts organelle advancement in stabilized (nonadvancing) cones. Addition of forskolin elevates cyclic AMP (cAMP) and allows microtubule extension into the periphery, accompanied by organelles (Fig. 4.3; Forscher et al. 1987). Similar destabilization follows treatment with cytochalasin, which prevents the addition of actin monomers to the actin filaments (Forscher and Smith 1988).

Consolidation of the axon behind the advancing growth cone may involve bundling of the microtubules by microtubule-associated proteins and collapse of the plasma membrane around the narrowed structure (Sabry et al. 1991).

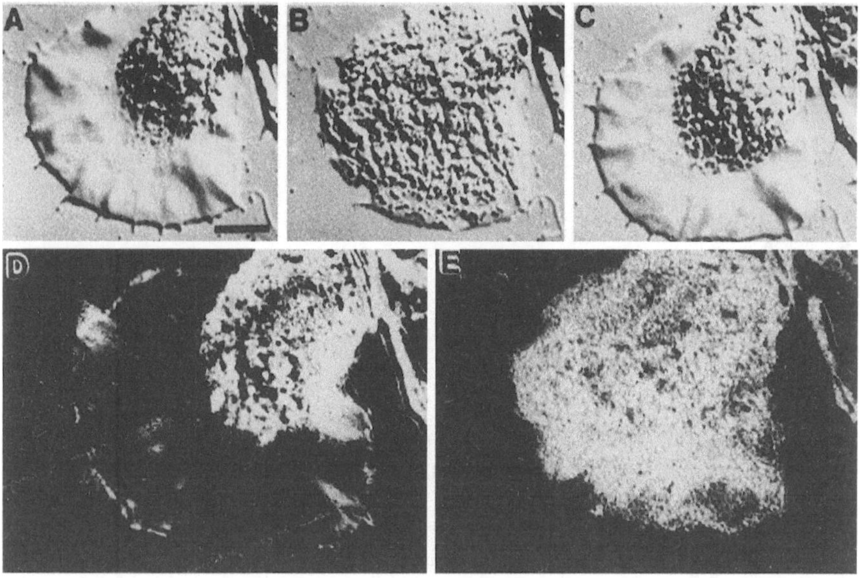

Fig. 4.3. Forskolin treatment induces extension of the central domain. **A** Video-enhanced differential interference contrast image of growth cone prior to treatment in Na-ASW control solution. Note clear demarcation between peripheral (P-) and central (C-) domains. **B** After 15 min of continuous exposure to Na-ASW containing 50 μM forskolin and 2 mm theophylline. **C** Recovery after Na-ASW wash. Note that the P – C domain interface returns to roughly the same position as in A, restoring the P C domain ratio observed in the pretreatment control. Scale *bar* = 5 μm. **D** RMSAD (root mean square average of differences) analysis illustrating movement under control conditions (as in **A** and **C**); 20 frames, 3.5-s interframe interval. **E** RMSAD analysis in the presence of forskolin/theophylline (as in **B**). (Forscher et al. 1987)

Growth cones advance spontaneously when a discrete region of the cone between neighboring filopodia develops bridging regions called veils (Fig. 4.4; Goldberg and Burmeister 1989). Membrane protrusion precedes engorgement, the movement of organelles into the expanded region by fast axoplasmic trans port along microtubules. The previous site of the central domain narrows and becomes part of the axon, dominated by microtubules (Fig. 4.4). A similar sequence is described for grasshopper growth cones, in which contact with a high-affinity substrate by a single filopodium can reorient the growth cone (Fig. 4.2; O'Connor et al. 1990). Studies of leech neurons growing on a patterned matrix (Sect. 4.9) have shown that elongation of the neurite is not restricted to the vicinity of the growth cone (Fromherz and Schaden 1994).

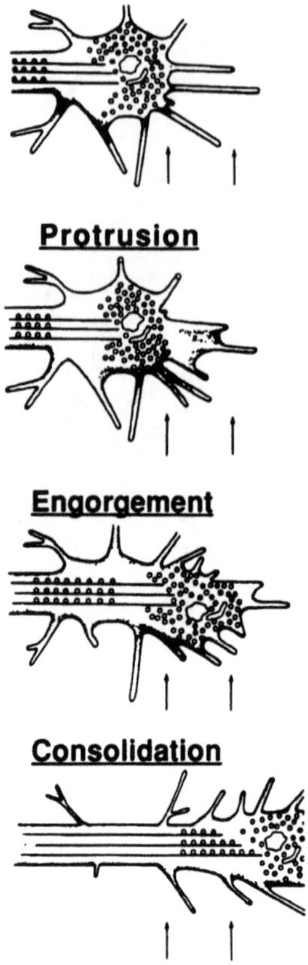

Fig. 4.4. Steps in the formation of a new segment of neurite, between the *arrows*. *Lines* represent microtubules. (Goldberg and Burmeister 1989)

4.3.3 Responses to the Environment

Growth cone elongation and the activity of motile filopodia are separable, individually regulated functions, although intracellular Ca^{2+} levels may be involved in both processes (Kater and Mattson 1988). Cytochalasin B disrupts filopodia and seriously interferes with pathfinding, but it does not inhibit growth per se of grasshopper pioneer neurons (Bentley and Toroian-Raymond 1986). Loss of navigability supports the idea that filopodia normally possess the sensory transduction elements of the growth cone. Similar navigational errors are made by embryonic *Xenopus* neurons treated with cytochalasin B (Chien et al. 1993).

The degree of elaboration of the growth cone is related to the complexity of the pathfinding task. Pioneer neurons are characterized by large growth cones with a large number of filopodia that display greater activity, compared with the neurons that follow them (LoPresti et al. 1973; Klose and Bentley 1989; Chitnis and Kuwada 1991; Kim et al. 1991). Pioneers often are competent to respond to multiple cues (Blair et al. 1987; Condic and Bentley 1989c).

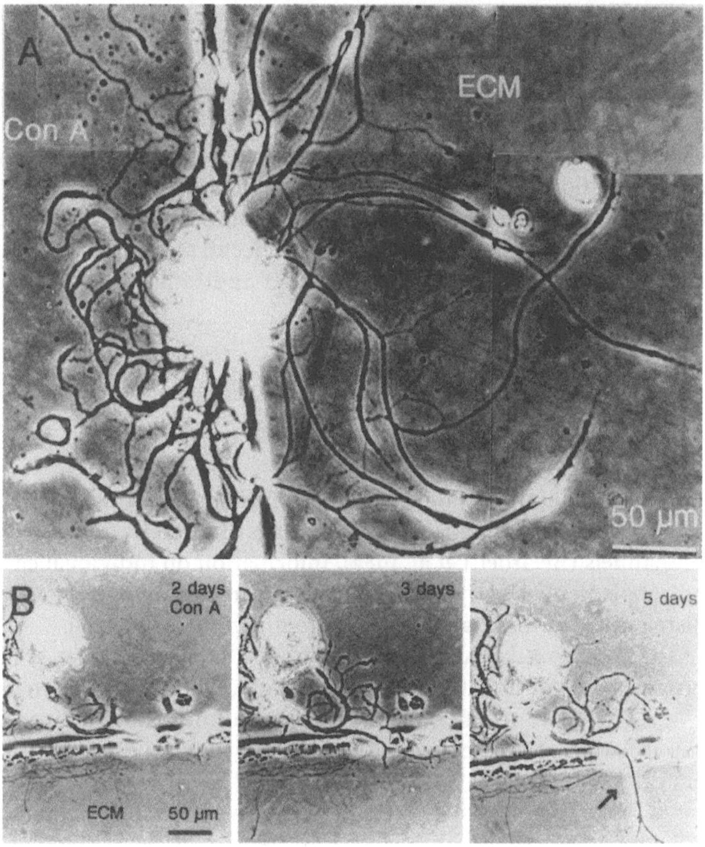

Fig. 4.5.A Neurite outgrowth of AP cells sprouting for 4 days in culture on a patterned substrate. On the *left side* of the border, the plate was coated with *Con A* and on the *right side* with *ECM*-laminin. **B** Individual branches of neurite (*arrow*) cross from one substrate (*Con A*) to the other (*ECM*-laminin). (Grumbacher-Reinert 1989)

The substrate influences the morphology of growth cones of cultured neurons. Neurites of leech neurons assume different morphologies when cultured on different substrates (Chiquet and Acklin 1986; Chiquet and Nicholls 1987; Masuda-Nakagawa and Wiedemann 1992). Neurons plated on the plant lectin concanavalin A (Con A) produce broad, short processes, whereas those grown on extracellular matrix containing laminin extracted from leech ganglion capsules extend processes approximately four times the distance of the processes grown on Con A, with only a third as many branches (Grumbacher-Reinert 1989). In contrast to the straight, relatively unbranched pattern seen on laminin, neurons grown on leech tenascin have profuse, curving neurites (Masuda-Nakagawa and Wiedemann 1992). Neurons grown at the juncture of two substrates produce distinct classes of neurites in response to the particular substratum upon which they extend (Fig. 4.5). (See Sect. 4.4.3 for a discussion of channel activity as a function of different substrates.)

4.4 Ion Channels in the Growth Cone

Receptors which govern the admission of ions represent one pathway for information flow from the environment to the cell interior (Spitzer 1991). Attention has focused on the distribution of Ca^{2+} currents in growth cones because so many functions, including cytoskeletonal alterations, responses to neurotransmitters, transmitter release, and membrane-substrate interactions, are expressed by changes in internal Ca^{2+} concentration (Hume et al. 1983; Delaney and Zucker 1990; Silver et al. 1990; Kater and Mills 1991).

4.4.1 Recording Channel Activity

There are several ways to examine channels in the growth cone.

1. Isolated membrane patch recordings allow detection and characterization of channels (Belardetti et al. 1987). Information on individual channel types can be supplemented by macroscopic current recordings that indicate the relative distribution of the channels in different regions of the cell.

2. Loose patch clamp sampling from a relatively larger area of membrane provides similar quantitative channel data (Nicholls and Hernandez 1989).

3. The anatomy of neurons grown in culture is often favorable for sucrose gap isolation of different regions of the cell so that ion substitutions and blockers can be employed (Gottman et al. 1991).

4. Voltage-sensitive dyes or Ca^{2+}-sensitive dyes such as Fura 2 and Arsenazo III can be used to visualize electrical events relative to entry of Ca^{2+} (Grinvald and Farber 1981; Cohan et al. 1987a; Ross et al. 1987; Bentley et al. 1991; Howes et al. 1991).

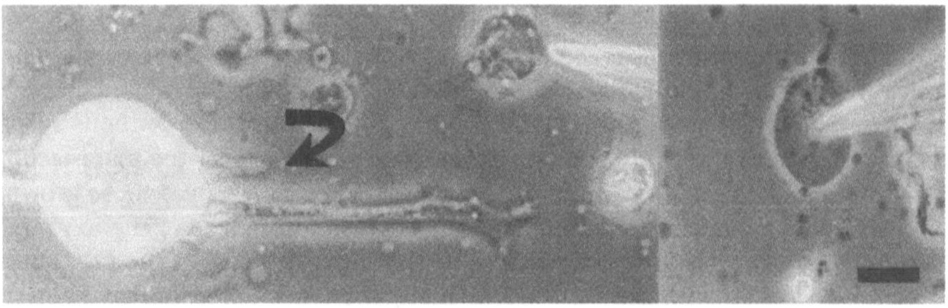

Fig. 4.6. Growth balls with attached patch electrodes. A growth ball is obtained by severing a growth cone from its neurite, whereupon the growth cone rolls up and the neurite retracts, as indicated by the *arrow* in *left frame*. *Right frame* shows a phase micrograph of a growth ball from another neuron. Neurons were plated 3 days prior to growth cone manipulation on glass coverslips coated with 10 µg/ml of either poly-L-lysine, concanavalin A, or *Helix aspersa* lectins. Cells were cultured in Leibovitz-15 medium with appropriate salt concentration. *Horizontal bar* = 10 µm (Marom and Dagan 1987)

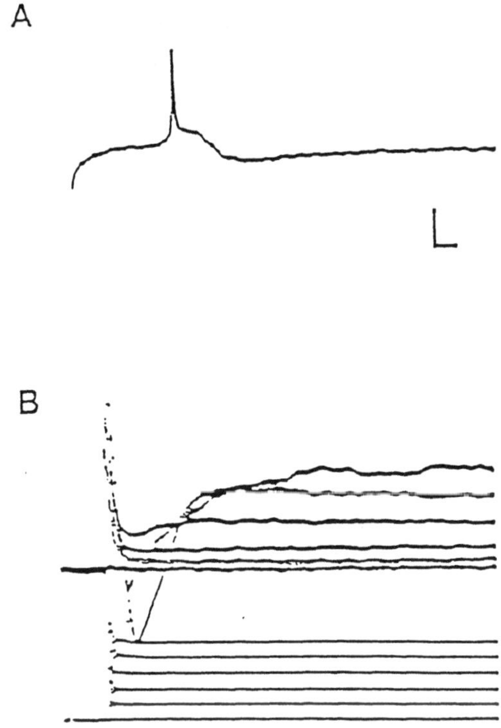

Fig. 4.7.A An action potential recorded from an isolated growth cone under current-clamp conditions. Records were obtained in the whole-cell configuration with a single-electrode voltage clamp amplifier. Calibration: 7 mV, 25 ms. **B** Voltage-clamp recordings of the 10 μm diameter growth ball shown in first frame of Fig. 4.6. An early inactivating inward current is seen, followed by a noninactivating outward current. Holding membrane potential: -70 mV. Calibration bar: 35 mV, 160 pA, 5 ms. (Marom and Dagan 1987).

5. The growth cone can be physically isolated from the cell by compressing the shaft with a glass microelectrode used as a microknife. This technique produces a rounded organelle (Fig. 4.6) that can be subjected to whole-cell patch clamp recordings (Fig. 4.7). Such "growth balls" isolated from growth cones of cultured *Helix* neurons exhibit voltage-dependent Ca^{2+} currents (Fig. 4.8; Marom and Dagan 1987).

6. Differential density centrifugation can render membrane vesicles from the growth cones for mass analysis of responses (Meiri and Burdick 1991).

4.4.2 Dynamics of Growth Cone Channel Distribution

We possess the techniques to determine the presence and relative numbers of different currents, but the growth cone can be a slippery subject to characterize (Spitzer 1991). The receptors and channels in its membrane reflect its condition and history.

Following axotomy, the properties of the membrane at the end of the axon initially reflect the nature of the membranes that seal the axon (Chap. 3). After the cockroach giant axon becomes sealed, the membrane initially supports a Na^+-dependent action potential which is replaced by a Ca^{2+}-dependent action potential (Meiri et al. 1981).

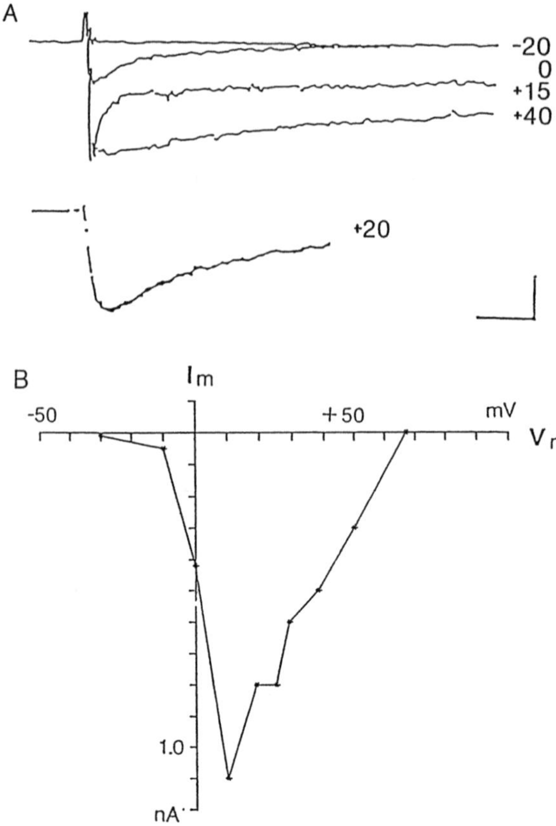

Fig. 4.8. Calcium currents from a growth ball. **A** Currents recorded from a growth ball under whole-cell patch-clamp configuration. *Upper four traces* were obtained with a patch clamp amplifier, while the *bottom trace* was obtained with a single-electrode voltage-clamp amplifier. Bath solution contained: 10 mM $CaCl_2$, 5 mM $MgCl_2$, 5 mM 4-AP, 20 mM TRIS-HCl, 35 mM TEA-Cl, 50 μm TTX, pH 7.4. Pipette solution contained 120 mM CsCl, 5 mM EGTA, 5 mM HEPES, pH 7.4. Holding potential: -50 mV. Calibration of *upper four traces*: 40 pA, 16 ms; *bottom trace*: 400 pA, 8 ms. **B** Voltage-current curve of a calcium current recorded from the growth ball shown in the right frame of Fig. 4.6. (Marom and Dagan 1987)

Marom and Dagan (1987) used TTX to reveal Ca^{2+} currents in *Helix* growth cones (Fig. 4.8) and both stretch-activated and stretch-inactivated channels have been detected in *Lymnaea* growth cones (Sigurdson and Morris 1989). In *Aplysia*, the growth cones from abdominal ganglion left upper quadrant neurons contain voltage-dependent Na^+ and K^+ channels, but no Ca^{2+} currents of the sort recorded in the soma (Belardetti et al. 1987). In the light of the effects of substrate on leech neurons (Ross et al. 1988; Sect. 4.4.3), different culture conditions may yield different observations.

4.4.3 Substrate Effects on Growth Cone Channels

The type of substrate may influence which channels are present in growth cones. In leech Retzius cells grown on Con A, the appearance of the cells is different from the growth pattern on extracellular matrix (Fig. 4.9). Large Ca^{2+} signals are recorded in the soma and even larger signals in the initial segments of

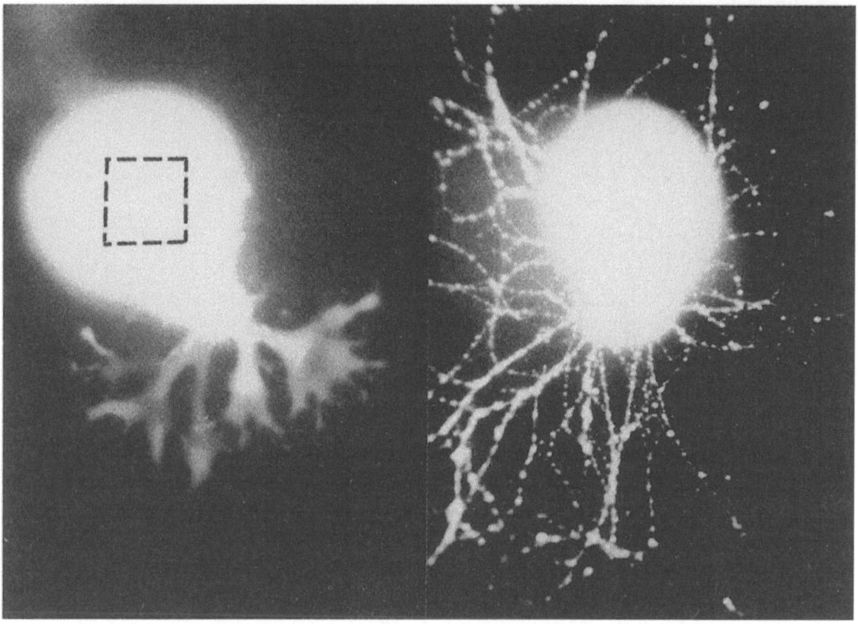

Fig. 4.9. Comparison of arborizations of Retzius cells (5 days) plated on Con A (*left*) and on ECM leech laminin (*right*). Note the broad flat processes on Con A. A field 40 μm × 40 μm is shown by *dashed lines*. Far less membrane area is available in the processes on the laminin for recording changes in absorbance. Cells were injected with Lucifer yellow and arsenazo III. Slight beading of the processes on the laminin extract is the result of photodynamic damage during exposure of Lucifer yellow to UV light for photography at the end of the experiment. (Ross et al. 1988)

cells grown on Con A, but no Ca^{2+} entry is recorded in distal processes and growth cones (Grumbacher-Reinert and Nicholls 1992). In contrast, Retzius cells grown on leech laminin have strong Ca^{2+} signals in even the finest processes (Fig. 4.10). This difference is not so marked in other identified neurons, but a similar trend is detectable (Ross et al. 1987, 1988). From such studies, one begins to understand how features of the environment can shape the growing neuron's morphology and responsiveness.

The leech neurons exhibited a different Ca^{2+} channel distribution pattern on natural and artificial (Con A) substrates. In contrast, the microenvironments provided in experiments with regenerating cockroach giant neurons did not influence Ca^{2+} channel distribution, which appears to be regulated by intrinsic factors (Spira et al. 1987). The voltage-regulated Ca^{2+} channels are found along the regenerating axonal segment as well as at the growing tip; this pattern was seen both in the normal ganglionic environment and in ganglia subjected to freezing and thawing, a procedure which altered the neurite's morphology (Spira et al. 1987).

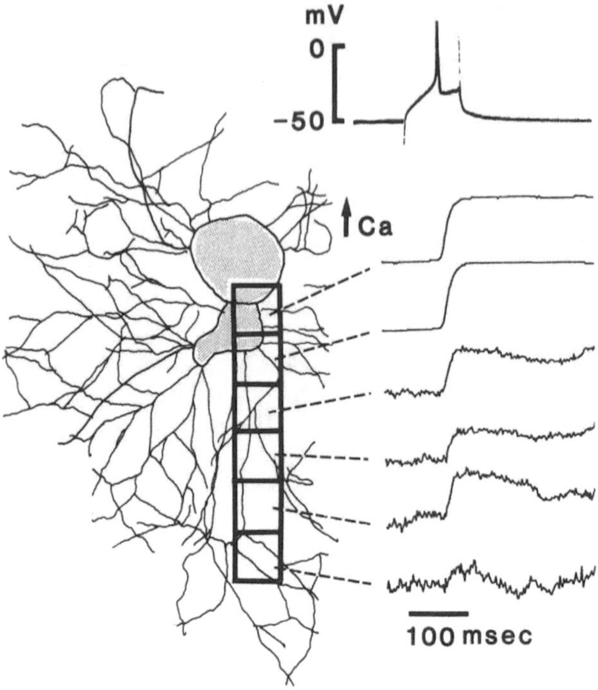

Fig. 4.10. Ca^{2+} signals from a Retzius cell grown on leech laminin (right cell in Fig. 4.9). The cell was stimulated 195 times through the microelectrode. The electrical recording shows one of the action potentials that is typical for Retzius cells in the ganglion or in culture. The tracings on the *left* show the outline of the cell body, the initial segment, and the processes. The positions of 6 of the array of 96 photodetectors are shown. Each detector registered changes in absorbance corresponding to an increase in Ca^{2+} in an area of 40 μm^2. The area of membrane in the *lower boxes* is at least an order of magnitude smaller that that for a cell on Con A, such as the cell on the left in Fig. 4.9, in which no Ca^{2+} signals were recorded from such processes. Although the actual amplitudes of the Ca^{2+} transient signals are smaller for boxes over the small processes, unambiguous time-locked changes were evident. (In this figure, all records have been adjusted to enhance the signals; therefore, the scale factor for each trace differs.) Other detectors over fine processes (not shown) also exhibited clear changes in Ca^{2+}. (Ross et al. 1988)

4.4.4 Channel Expression Reflects Growth State

Another factor known to influence the presence and activity of channels is the growth state, i.e, whether a growth cone is in a growing or a stable condition. The state of the neuron and its growth cones are influenced by the length of time in culture, intrinsic factors, and interactions with the environment (Grumbacher-Reinert and Nicholls 1992).

Neurons from the snail *Helisoma* undergo a predictable transition from an actively growing state to a stable morphology after several days in culture. The

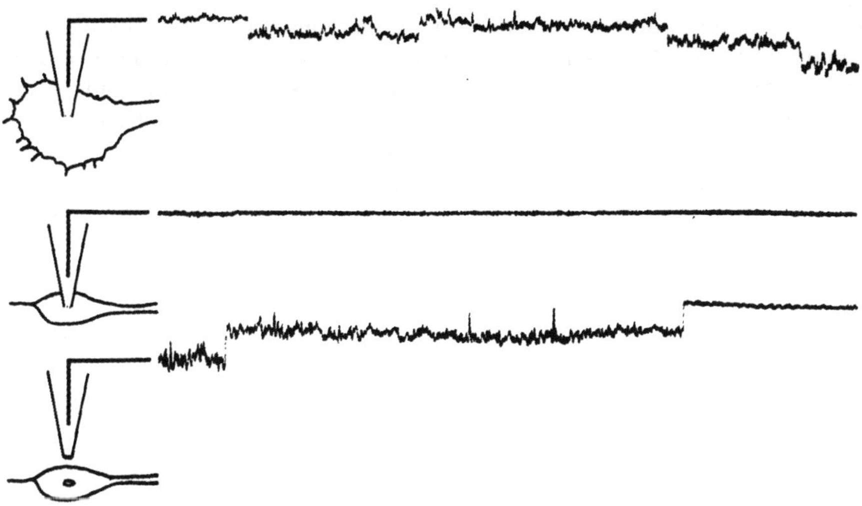

Fig. 4.11. Differences in single channel activity of growing as compared to stable growth cones. Drawings on the *left* indicate the type of patch and the growth state of the growth cone. Channel activity that is present in cell-attached patches of growing growth cones (*top trace*) is absent from cell-attached patches of stable growth cones (*middle trace*). The onset of channel activity in patches from stable growth cones after the pipette has been removed from the cell (*bottom trace*) indicates that channels are still present in stable growth cone membrane but were previously silent. Middle and bottom traces are from the same membrane patch. A holding potential of -50 mv was applied to the pipette for both cell-attached records. The excised patch was held at a potential of -100 mv. Assuming a resting membrane potential of -50 mv for these neurons, the patches in these three records appear at similar membrane potentials. Calibration: 6 pA, 600 ms (Cohan et al. 1985)

characteristics of the channels in growing and nongrowing growth cones have been compared in patch clamp recordings (Cohan et al. 1985). An ion channel with a conductance of 70 pS is present in both membranes, but in actively growing cones the channel is spontaneously active, whereas in stabilized growth cones the channel is inactivated. In stable membrane the presence of the channel can be unmasked by excising a membrane patch during a recording (Fig. 4.11). Cytoplasmic systems other than (or in addition to) the internal Ca^{2+} concentration regulate the activation state of the channel.

4.4.5 Electrical Synapses and Intercellular Communication

In addition to considering channels that bridge the membrane from the intracellular to the extracellular environment, it is also useful to consider the role of channels that form to allow communication between cells. The channel proteins (connexins) that form gap junctions provide contact sites for cell-cell communication in developing and regenerating nervous systems.

Transient connections are often important in nervous system and muscular system development (Anderson et al. 1980; Kirk and Govind 1992). Studies of *Drosophila* eye development (Tomlinson and Ready 1987; Cagan and Ready 1989) and *Drosophila* genetic mosaics have revealed the role of information transfer in retinal cell fate determination (Cagan and Zipursky 1992). Channels in *Drosophila* allow the transfer of negatively charged and neutral particles up to a molecular weight of 1000 (Ryerse 1989). The exchange of either ions or macromolecules can alter developmental processes of linked cells. In the visual system of the crustacean *Daphnia*, undifferentiated cells become neurons following growth cone contact (LoPresti et al. 1973, 1974). During development of the abdominal ganglion in *Aplysia*, axonogenesis is initiated following the formation of axosomatic synapses (Schacher et al. 1979).

Intracellular Ca^{2+} levels are adjusted via electrical contacts. In the grasshopper *Schistocerca*, transient electrical connections between pioneer neurons and preaxonogenesis neurons allow Ca^{2+} transfer: prior to axonogenesis, grasshopper neurons exhibit Ca^{2+} levels in the 65 nM range, and after transient contact, outgrowth is initiated and the concentrations are above 100 nM (Bentley et al. 1991). Synaptic currents may be the stimulus that brings the Ca^{2+} concentration into a range required for the neurons to sprout in culture (Sect. 4.4.9.2), but other factors may also be involved.

Electrical synapses are observed in many invertebrate in situ and in vitro culture studies. Certain combinations of leech neurons form rectifying junctions in culture and others form nonrectifying junctions (Nicholls 1986). In pairs of cultured leech neurons tested at one week, junctions between two L cells (longitudinal) or two AP cells (anterior pagoda) are nonrectifying, but earlier, the junctions are weak and polarized, favoring current spread in one direction (Davis 1989).

4.4.6 Developmental Regulation of Channel Expression

Developmental studies show that cells in the early embryo possess a variety of voltage-gated channels, and channel expression is regulated by intrinsic as well as extrinsic factors (Moody et al. 1991). For instance, in the ascidian embryo, an event which distinguishes epidermal from neural differentiation is suppression of the inward-rectifier K^+ channel (Okamura and Takahashi 1993) and later interactions affecting channels are described for this embryo (Okamura et al. 1993).

Likewise, a variety of channels appear during the life of the growth cone (Spitzer 1991). Voltage sensitive Na^+ and Ca^{2+} channels may be expressed early in development, with broad action potentials dominated by Ca^{2+} entry. Spontaneous activity is seen in cultured amphibian spinal neurons during the Ca^{2+}-dominated stage of development (Holliday and Spitzer 1990), and Ca^{2+} channels may underlie spontaneous firing in other neurons. Factors that influence the growth cone's membrane potential regulate Ca^{2+} entry through voltage-gated channels.

Expression of K^+ channels, which depress excitability, abbreviate the action potential, and limit Ca^{2+} entry, is also developmentally regulated. In amphibian spinal neurons, expression of the delayed rectifier current is linked to Ca^{2+} entry at an earlier stage of development (Desarmenien and Spitzer 1991). In *Drosophila* neuroblasts, K^+ channels and Ca^{2+} channels are present early; increased excitability coincides with the addition of Na^+ channels (Wu et al. 1983). *Drosophila* mutants provide insights into the role of particular channel proteins in growth cone function (Wu 1988).

Transcription can be regulated by Ca^{2+} (Dash et al. 1991; Sheng et al. 1991). The concentration of Ca^{2+} can be higher in the nucleus than in the cytoplasm. Ca^{2+} entering across the plasma membrane is coupled to the Ca^{2+}-stimulated release of Ca^{2+} from the endoplasmic reticulum; this is required for maturation of the delayed rectifier potassium current in *Xenopus* spinal neurons (Desarmenien and Spitzer 1991).

4.4.7 Sensitivity to Electrical Fields

The presence of electric fields in developing vertebrate embryos and regenerating vertebrate limbs has been recognized for many years. Mechanisms of neural responses to electric fields include facilitation of slow axonal transport in rat spinal cord (McLane et al. 1994) and stimulation of Ca^{2+} entry through voltage-gated channels in cultured *Xenopus* neurons (Stewart and McCaig 1994). Little is known about the possible presence of electric fields in invertebrates and their roles in regeneration. On the cellular level, field effects certainly exist, as discussed in connection with the entry of ions after axotomy (Chap. 3, Sect. 3.3). The neurites projecting from leech neurons growing in culture have pronounced gradients of Na^+ channels, with the highest sodium currents recorded at the tips (Nicholls and Hernandez 1989).

Focused electric fields promote reconnection of giant axons in the earthworm (Todorov et al. 1992; Sect. 3.4). The success of neurite orientation to either surviving axonal segments (in instances where fusion is possible) or degenerating distal stumps (where cues to correct reinnervation of targets exist) could involve responses to naturally occurring electric fields.

4.4.8 Roles of Electrical Activity in Growing Neurites

Spontaneous activity is seen in many neurons at various stages of development. Its appearance may reflect an intrinsic time table, as in the spontaneous activity seen in embryonic insect neurons after 3 weeks in culture (Lees and Beadle 1991). The activity often plays a role in differentiation, e.g., in amphibian spinal neurons and vertebrate and insect visual systems (Constantine-Paton et al. 1990; Holliday and Spitzer 1990; Meinertzhagen 1993; Mimura 1993; Collett 1994).

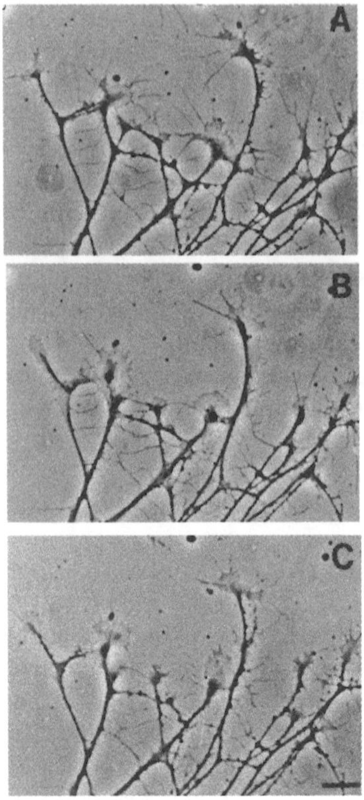

Fig. 4.12. Comparison of growth cone morphology during control, stimulation, and recovery periods. **A** During the control period, growth cones of B4 in this field display numerous filopodia extending from their lamellipodia. **B** During stimulation (4/s) note the significant decrease in the number of filopodia per growth cone. **C** During the recovery period, filopodia appear again on the growth cones. Calibration *bar* = 20 μm. (Cohan 1990)

Stomatogastric ganglion neurons of the spiny lobster that are driven by rhythmic inhibitory inputs are silent when pharmacologically isolated in vivo or when physically isolated in vitro. However, they become capable of generating endogenous bursts after 3–4 days in culture (Turrigiano et al. 1994). The transition involves increasing Ca^{2+} current and decreasing outward current. Synaptic inputs may normally regulate the membrane properties of the cells, when they are part of a rhythmic circuit. The transition observed in these cells would potentially restore rhythmic behavior if some members of the circuit were lost.

Action potentials propagate into the processes and growth cones of regenerating neurons of some animals (Ross et al. 1988), and isolated growth cones can generate action potentials, so the voltage-sensitive channels characteristic of a particular neuron may be present in its growth cones. What functions might be served by this? The propagation of action potentials into growth cones could increase neurotransmitter release if the excitation-secretion mechanism is functional. Also, electrical events spreading from the axon into a growth cone could regulate growth by activating second messenger systems responsive to the entry of Ca^{2+}. Depending on the baseline Ca^{2+} concentration, elevation could either promote growth or stabilize morphology (Cohan and Kater 1986).

An inhibitory effect of depolarization on the growth cone is demonstrated by attaching extracellular patch pipettes to the soma of cultured *Helisoma* neurons (intracellular electrodes often stop elongation) and passing current to evoke action potentials (Cohan and Kater 1986). Action potential frequencies well below the normal activity range for buccal neuron B19 halt growth cone advance in a reversible manner. Associated changes in the cell's morphology include retraction of filopodia and lamellopodia (Fig. 4.12). Neuron B4 is more sensitive to stimulation and better able to rebound from particular stimulus frequencies with accelerated growth than B19 (Cohan 1990). Similar responses are reported for leech neurons (Grumbacher-Reinert and Nicholls 1992).

The growth rate of phasic and tonic crustacean motoneurons in explant cultures is correlated with levels of spontaneous activity: tonic motoneurons show spontaneous activity for up to 6 days in culture, and grow more slowly and branch less freqently than the silent phasic motoneurons (Egid and Lnenicka 1993). Depolarization with elevated external K^+ causes retraction of growth cones in both types of motoneurons, but tonic cells recover whereas phasic cells degenerate (Egid and Lnenicka 1994).

4.4.9 Roles of Calcium

4.4.9.1 Calcium and the Cytoskeleton

Cytoskeletal regulatory systems are important targets for Ca^{2+} second-messenger systems. Changes in growth cone structure, detachment or attachment to the substratum, and extension or retraction all involve polymerization or depolymerization of microtubules, microfilaments and membrane-associated skeletal elements. Different portions of the cytoskeleton have different sensitivities to a change in internal Ca^{2+}, so that graded responses can be produced. For instance, a slight elevation in intracellular Ca^{2+} will destabilize actin filaments while leaving microtubules intact, and lamellipodial actin and filopodial actin can be differentially regulated (Lankford and Letourneau 1989). Alpha actinin exists in Ca^{2+}-sensitive and Ca^{2+}-insensitive forms, and the ratio of these determines the sensitivity of the filaments in different regions of the cell.

Microtubulin associated proteins (MAPs) stabilize the structure of microtubules in a reaction that is subject to competitive inhibition by Ca^{2+}-calmodulin (Bamburg et al. 1986; Lin et al. 1989; Gordon-Weeks and Mansfield 1990). MAPs may also play a role in linking microtubules and actin filaments in the lamellipodia. The tau family is structurally related to MAP proteins, and both families promote microtubule assembly and stability in vertebrate neurons. Different isoforms of MAP2 are characteristic of different stages of development (Goedert et al. 1991).

4.4.9.2 Sprouting and Collapsing

In each *Helisoma* neuron there is a level of internal Ca^{2+} concentration at which growth is optimized (around 130 nM), and higher or lower levels inhibit sprouting (e.g., 56 nM is recorded in stablilized growth cones)(Kater and Mills 1991). *Helisoma* neurons normally require a conditioned medium or other promotors to initiate sprouting, but this requirement can be overcome by adding the channel blocker La^{3+}, which reduces intracellular Ca^{2+} to the growth-permissive range (Fig. 4.13; Mattson and Kater 1987; Mattson et al. 1988b). Sprouting in an otherwise permissive conditioned medium can be blocked by elevating internal Ca^{2+} with the Ca^{2+} ionophore A23187 (Mattson et al. 1988b).

Growth cone size and filopodial length increase at decision points, where the environment is sampled more widely (Taghert et al. 1982; Raper et al. 1983; Tosney and Landmesser 1985; Eisen et al. 1986). Regulation of filopodial length by intracellular Ca^{2+} has been studied in whole growth cones and in local regions within the growth cone (Rehder and Kater 1992; Davenport and Kater 1992). Experimental treatments which transiently elevate intracellular Ca^{2+} cause corresponding increases in the length of filopodia. The increases in filopodial length are transient, lasting for approximately 10 min, and the subsequent reduction in filopodial length occurs even when the Ca^{2+} homeostatic mechanisms are unable to drive the intracellular Ca^{2+} concentration back to the pretreatment range (Rehder and Kater 1992). When intracellular Ca^{2+} concentration is increased in a restricted portion of the growth cone by focal depolarizing current, the major effect is localized extension of filopodia (Davenport and Kater 1992).

Electrical stimulation causes growth cone collapse in growing leech neurites and the reinitiation of growth in cells that have ceased to grow (Grumbacher-Reinert and Nicholls 1992). Ca^{2+} entry is probably responsible for the changes because when the cells are grown in elevated Mg^{2+} solutions, the effect of stimulation is reduced.

In the peptide secreting neurons of the crustacean X-organ, two outgrowth patterns, veiling and branching, characterize two distinct cell types. Both types of cells grow with a wider range of internal Ca^{2+} than is seen in most neurons (50–300 nM). The two cell types maintain similar basal levels of internal Ca^{2+} under identical culture conditons. The voltage-dependent Ca^{2+} current is more prominent in the veiling cells, and these cells are more sensitive to the Ca^{2+}-blocking effects of Cd^{2+} than the branching cells (Graf and Cooke 1994). Addition of TTX had no effect on the growth of either cell type.

The growth cones of dorsal root ganglion neurons of fetal mice also respond to electrical and particularly phasic stimulation by collapsing and retracting (Fields et al. 1990). TTX blocks this response, which suggests that it depends upon voltage-gated Na^{+} channels. In contrast to the findings with snail neurons, collapse is not associated with change in intracellular Ca^{2+} concentration (Ivins et al. 1991).

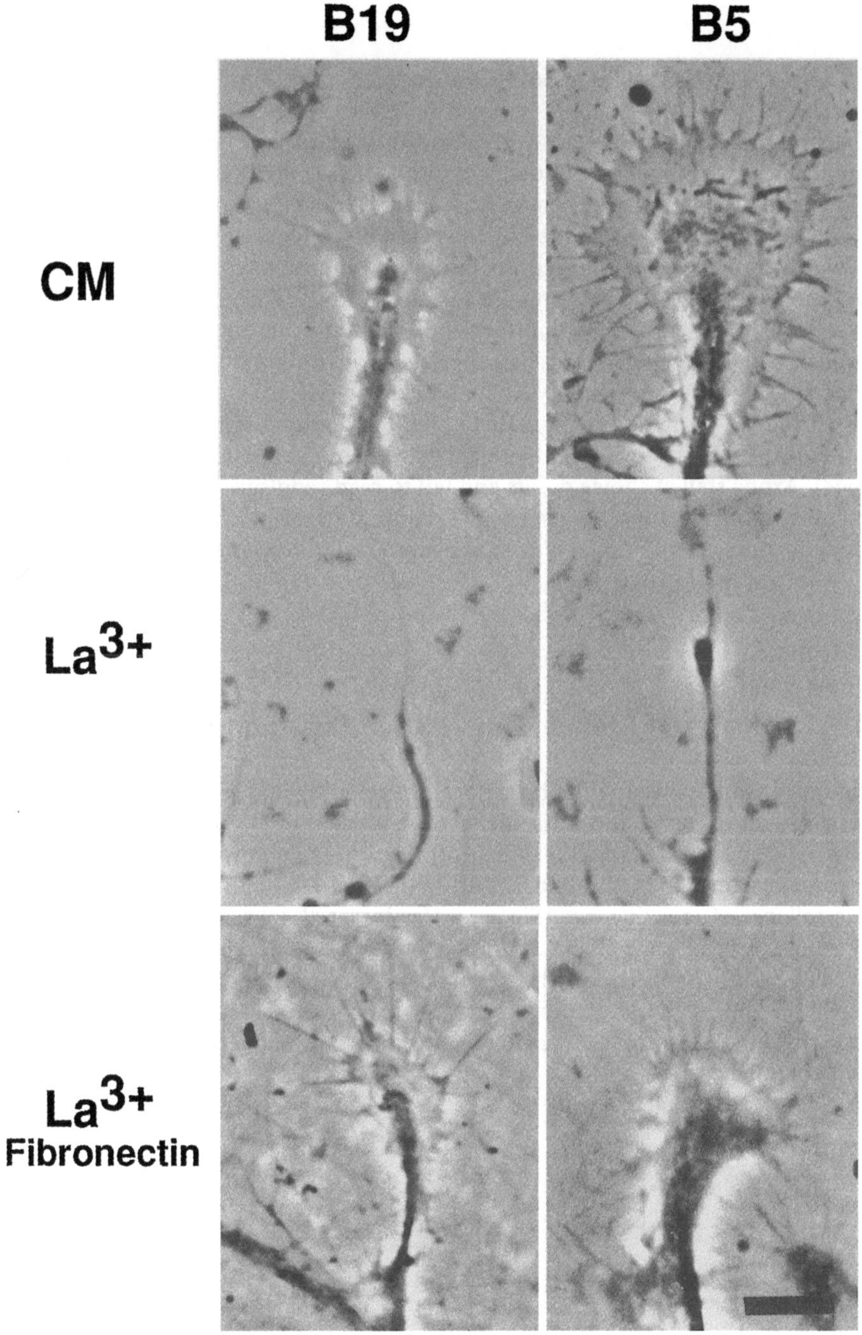

Fig. 4.13. Neuron-specific growth cone morphologies induced by fibronectin and CM. Growth cones of neurons B5 and B19, respectively, growing in response to either *CM* alone, *La^{3+}* (10 μM) alone, or *La^{3+}-fibronectin* (a substrate coated with 10 μg/ml of fibronectin). The micrographs are representative of 14–30 growth cones examined. Note that in the absence of a specific CM or fibronectin substrate, growth cone motility is greatly reduced (*upper* and *lower* vs. *middle*). Also note that growth cones of neurons B19 and B5 retain distinguishing features (i.e., filopodial numbers and lamellipodial span) when grown on a fibronectin substrate in the absence of CM (*upper* vs. *lower*). Calibration *bar* = 10 μm. (Mattson et al. 1988b)

4.4.9.3 Ca^{2+} Effects Mediated by Second Messengers

Calcium can interact directly with some intracellular targets, but more commonly it acts through calmodulin or other messenger systems (Gnegy 1993; Hinrichsen 1993). Evidence that at least some of the effects of Ca^{2+} are mediated through calmodulin is provided by the calmodulin blocker, CGS 9343B. In *Helisoma* neuron B19, CGS blocked the inhibition of growth cone elongation without blocking the electrical response to transmitters or the entry of Ca^{2+} (Polak et al. 1991). The most common form of calmodulin in neurites is calmodulin protein kinase II, which, among other reactions, phosphorylates cytoskeletal proteins, including tubulins, MAP-2 and synapsin I.

If the Ca^{2+}-calmodulin signaling system is disrupted in *Drosophila* growth cones, errors in extension and guidance are seen (Van Berkum and Goodman 1995). Axons stall at choice points and fail to read repulsive cues in the environment that normally restrict crossing of the midline. Repulsive cues in *Helisoma* growth cones lead to Ca^{2+}-mediated collapse (Bandtlow et al. 1993).

4.5 Responses of Growth Cones to Neurotransmitters

Neurotransmitters shape the behavior of developing and regenerating neurons (McCobb et al. 1988a and b; Lauder 1993; Chiasson et al. 1994). Kater and colleagues demonstrated the role of neurotransmitters in growth cones of cultured *Helisoma* neurons. In some identified neurons, bath-applied serotonin causes the collapse of growth cones (Fig. 4.14; Haydon et al. 1987; McCobb et al. 1988a, b; Murrain et al. 1990). The activation of serotonin receptors opens voltage-sensitive Ca^{2+} channels and elevates internal Ca^{2+}, leading to growth cone collapse (Cohan et al. 1987a). Approximately half of the buccal ganglion neurons examined are insensitive to serotonin, and of those, some collapse in response to dopamine.

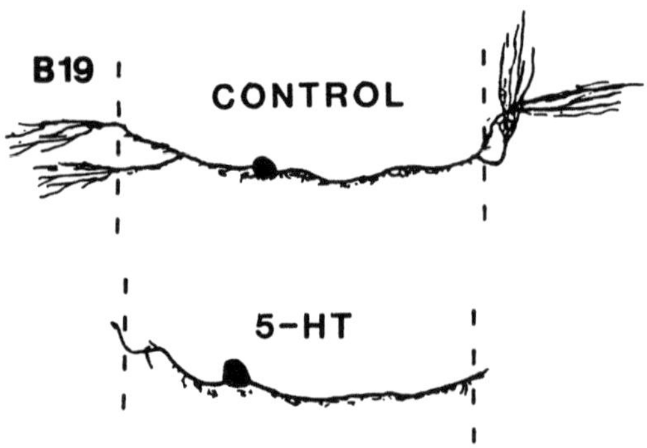

Fig. 4.14 Tracings of two regenerating buccal neurons 19. The neuron at the *top* was incubated in a control medium, and the neuron at the *bottom* was incubated in a medium containing 5×10^{-5} M 5-HT. The *dotted lines* indicate points of axotomy. (Murrain et al. 1990)

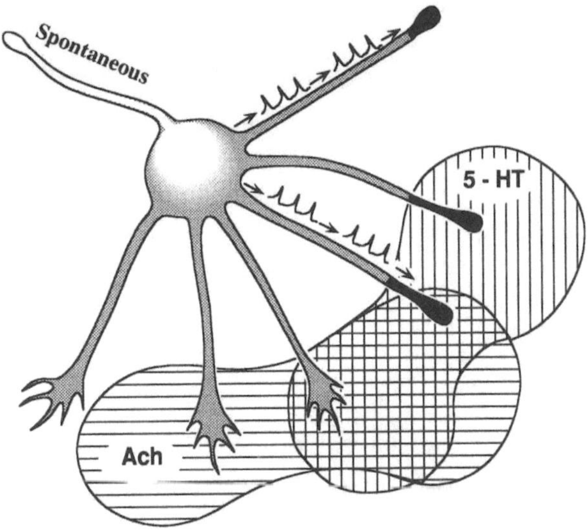

Fig. 4.15. The integration of multiple stimuli in the regulation of growth cone behavior. In the absence of particular stimuli, individual growth cones either continue outgrowth at a particular calcium level or spontaneously terminate outgrowth accompanied by decreases in intracellular calcium. The generation of action potentials abruptly inhibits neurite outgrowth (*darkened cone*), as does the presence of serotonin (*5-HT*), through a rise in intracellular calcium (*darkened growth cone*). The presence of *ACh* has no noticeable effects on growth cone motility or calcium levels, but can protect against 5-HT-induced inhibition of growth cone motility and rises in intracellular calcium levels. However, in the presence of both *ACh* and *5-HT*, the additional stimulus of action potentials raises intracellular calcium and results in an abrupt termination of outgrowth. This illustration, taken from results obtained with *Helisoma* neurons, exemplifies the kinds of conditions likely to be found in vivo, where multiple stimuli are integrated by the active neuronal growth cone. (Redrawn from Kater and Mills 1991)

The common feature of the responses to serotonin and dopamine is that both transmitters depolarize the cells that are sensitive to them. Transmitters that hyperpolarize the membrane of these cells, such as acetylcholine, can suppress growth cone collapse in the presence of serotonin by blocking the depolarization that opens voltage-sensitive Ca^{2+} channels (Fig. 4.15; Kater amd Mills 1991). Hippocampal pyramidal neurons respond to glutamate from dendritic growth cones in a similar Ca^{2+}-regulated response (Mattson et al. 1988a). Glia, as well as neurons, may be targets of the trophic effects of neurotransmitters (Lauder 1993).

Pathfinding is governed by both attraction and inhibition. Collapse of a growth cone represents an inhibited response that may occur over and over in the shaping of a neuron's final branching pattern. How does such a response interact with other events in development and regeneration? If the collapsed growth cone stops growing, the portion of the neuron that falls within the range of the inhibitory influence may be stabilized while other projections continue to

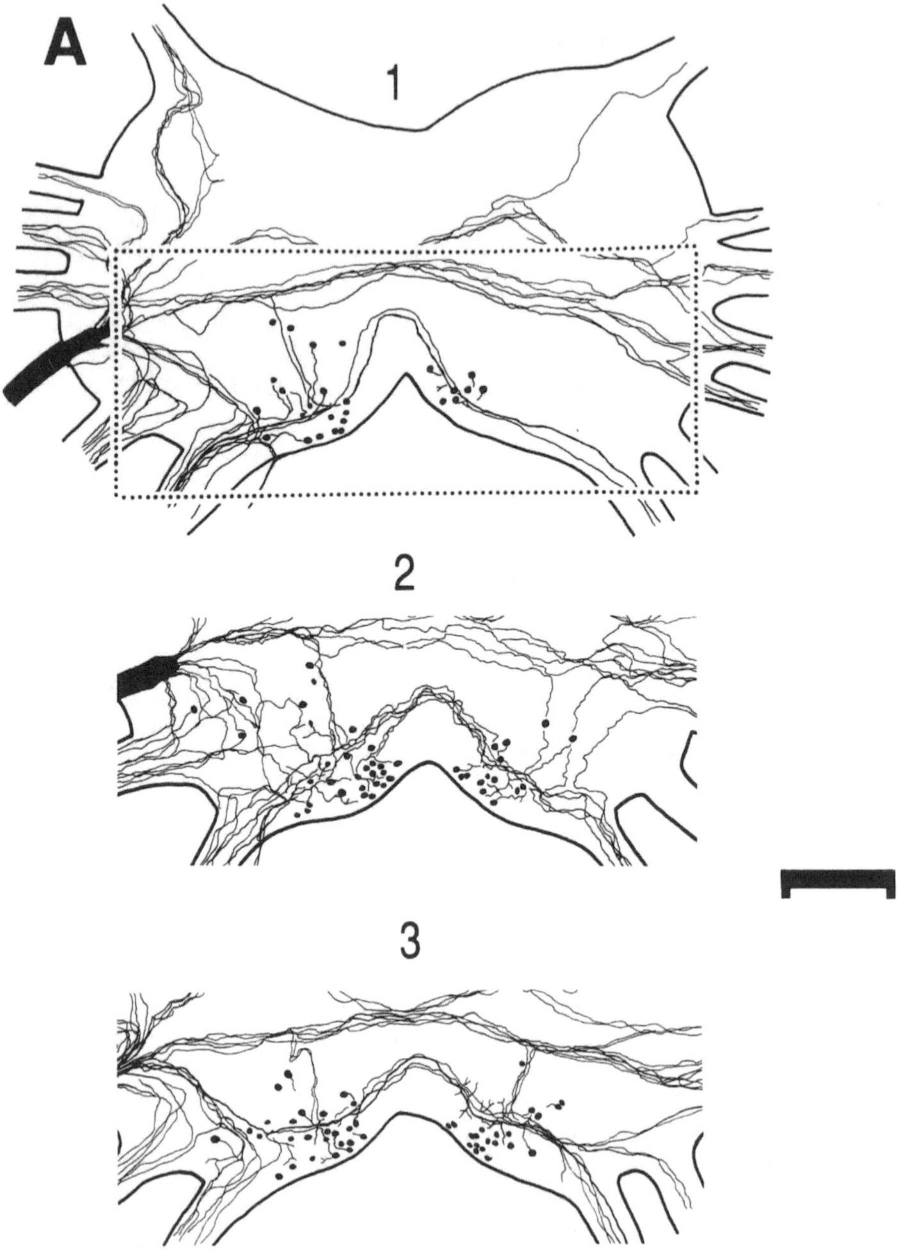

Fig. 4.16. *Camera lucida* tracings of typical cerebral ganglia following dye-fill of the cerebro-buccal connective (*left*). 1 Normal labeling on the dorsal surface. The *dotted outline* corresponds to areas which display supernumerary staining of neuronal elements, as illustrated below. 2 21 days after 5,7-dihydroxytryptamine treatment; 3 21 days after p-chlorophenylalanine treatment. Note especially the additional contralateral projections. Scale *bar* = 300 μm. (Baker et al. 1993)

expand (McCobb et al. 1988a,b). Guidance is provided if growth cones growing along a border between two substrates collapse whenever they move onto the less supportive substrate.

The relationship between collapse and synapse formation is not clear, but as long as a neurite is exposed to an appropriate transmitter, its structure is stabilized; when it ceases to receive that transmitter, it is free to explore new sources of the stabilizing transmitter. The importance of neurotransmitters in determining neurite morphology is evident from the alterations that follow serotonin blockage in the embryo (Goldberg et al. 1992). Moreover, in adult snails, suppression of serotonin release by depleters can lead to the sprouting of neurons in an otherwise intact nervous system (Fig. 4.16; Baker et al. 1993). The degeneration of a serotonin-releasing neuron, the metacerebral giant cell of the snail *Achatina*, leads to sprouting responses among uninjured neurons (Chiasson et al. 1994).

Developing gastropod neurons respond to serotonin in a more flexible manner than adult neurons (Goldberg et al. 1990, 1991). Serotonin stimulates the elongation of growth cones in embryonic *Helisoma* neurons when they are in a growth-arrested state, and inhibits growth when they are actively exploring. The fact that an individual neuron could alternate between these two responses suggests that a start-and-stop pattern of response to serotonin could shape growth without bringing about the relatively permanent stabilization that characterizes the adult cultured neuron response. This comparison between embryonic and adult neuron responses suggests that the give-and-take necessary for development has been restricted in adult neurons.

4.6 Differentiation Following Growth Cone Interactions

Environmental influences on developing neurons are exemplified in the differentiation of neurons that possess the same potential, but find themselves in different regions of an animal's body. Serially homologous neurons, such as those in insect sensilla, exhibit differences in morphology or connections related to segmental specializations (Palka 1993).

Retzius cells of the leech represent one of the best examples of extrinsically controlled development. The presence of reproductive organs in segments 5 and 6 of the leech alters the fate of the Retzius cells in these segments, where they innervate reproductive ducts rather than the body wall (Glover and Mason 1986). The synaptic connectivity and central morphology of these Retzius cells is also altered (Jellies et al. 1987; Loer et al. 1987; Loer and Kristan 1989; Wittenberg et al. 1990).

Experimental manipulations and examples of aberrant development have established that all Retzius cells have the capacity to respond to the presence of reproductive tissue or innervate the body wall (Macagno et al. 1986; Loer et al. 1987; Loer and Kristan 1989; French and Kristan 1992; Stent et al. 1992). When Retzius growth cones contact mesenchyme surrounding the reproductive tubes, they enlarge and project filopodia; other portions of the neuron then assume the densely branched innervation pattern characteristic of the reproductive segments (French et al. 1992).

4.7 Receptors and Molecular Cues in the Environment

4.7.1 Functions of External Signals

Developing invertebrate neurons depend upon many cues to direct their growth to intermediate and ultimate targets (reviewed by Whitington 1993). These cues include cellular and substrate-bound factors, and redundant cues may be provided. For instance, both the basal lamina and individual epithelial cells provide information for grasshopper pioneer growth cones (Condic and Bentley 1989a-c). Regeneration involves many of the same interactions that originally shaped the nervous system during development, and conservation of the molecular cues across phyla has aided their identification and characterization (Goodman 1994; Travis 1994).

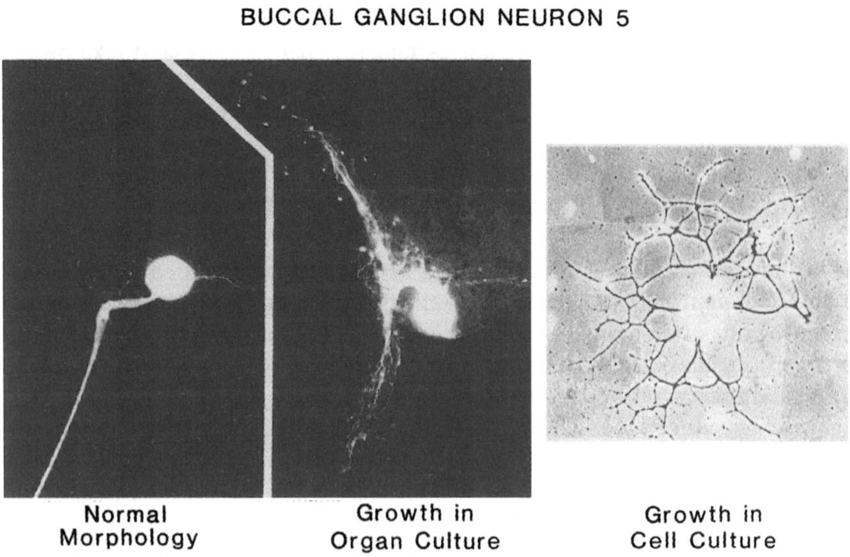

Fig. 4.17. Examples of three different environmental conditions in which a given neuron (in this case buccal ganglion neuron 5) can be experimentally examined and manipulated. Within normal ganglia (*left*) one can readily examine both physiological and morphological aspects of a normal neuron by standard intracellular microelectrode recording and dye-injection techniques. Similar methods, using Lucifer yellow staining, can be used in axotomized neurons within normal buccal ganglia either in situ or in organ culture. Such neurons grow profusely and form definable new connections (*middle*). Any given identified neuronal cell body can be removed from a normal ganglion and placed in cell culture (*right*) where again both physiological and morphological aspects of this neuron can be examined, but in this case, in the most defined of environmental conditions. (Kater 1985)

Both repulsive and attractive signals can be supplied on the surface of cells, in the extracellular matrix, or by diffusion from potential targets (Patterson 1988; Lander 1989; Halpern et al. 1991). When receptors in the neuron membrane bind signal molecules, growth cones respond following activation of intracellular cascades, such as the phosphatidylinositol pathway (Chang et al. 1992; Strittmatter and Fishman 1991; Atashi et al. 1992). The role of transient cellular targets in insect neural development has been studied intensively (e.g., Lakes-Harlan and Pollack 1993), but similar situations exist in many animals. Intermediate targets can include neurons, glia, muscle cells, and epithelia or associated basal lamina (Whitington 1993). Palka et al. (1992) suggest similarities between the utilization of guidepost cells and dependence on groups of cells to guide axon growth.

Kater (1985) illustrated the restricting role of the environment by comparing the anatomy of *Helisoma* buccal neuron 5 in the control condition, in an organ cultured ganglion, and in tissue culture (Fig. 4.17). In the control condition, the neuron has a relatively simple structure; the neuron regenerating within the ganglionic milieu branches more widely but similarities to the control suggest that cues within the ganglion influence its growth pattern. Neurons implanted into ganglia are restricted in the same manner (Zhang and Nicholls 1983; Syed et al. 1992a). In contrast, the isolated neuron projects extensively and uniformly over the culture substrate (Fig. 4.17).

Evidence that cues present in the ganglion influence the morphology of regenerating neurons was provided by comparing the typical pattern of cockroach giant axons regenerating through a ganglion with the growth seen when the ganglion had been subjected to freezing and thawing prior to entry of the axon (Spira et al. 1987). In the control situation, the axon has a large diameter, straight, smooth profiles in the connectives and a few fine branches in the ganglion. Regenerating axons follow this general pattern but exhibit many more projections in the connectives and ganglion. However, projections into ganglia that have been frozen do not branch at all, indicating that some cue has been destroyed by that procedure. The distribution of ACh receptors on the regenerating neurites is also disrupted by freezing the ganglion.

4.7.2 Binding and Recognition Systems

4.7.2.1 Introduction

The milieu through which growth cones project is a network, approximately 100 nm thick, consisting of proteins released from cells and the glycoproteins that extend from cell membranes. In favorable circumstances, constituents of the extracellular matrix can be extracted from the connective tissue capsule surrounding the CNS and nerves.

Adhesion systems have been described in a number of recent studies and reviews (Jessell 1988; Rutishauser et al. 1988; Lander 1989; Sanes 1989; Jessell et al. 1990; Landmesser et al. 1990; Schachner 1991; Venstrom and Reichardt 1993). Cell-cell interactions and cell-extracellular matrix interactions are both important, but do not represent a way to categorize the molecular families, because bound and unbound forms of the same molecules are sometimes encountered. Two categories of binding pattern are recognized: homophilic interactions involve binding between the same type of molecules (usually present on membranes of different cells) and heterophilic binding involves two different types of molecules.

Substrate and cell surface molecules have sometimes been named for their first identified action, such as collapsin, connectin or neural cell adhesion molecule, but, as is true for neurotransmitters, the effect of binding is determined by the particular receptor and the particular second messenger cascade that is subsequently activated. The same molecules can attract or repel, stimulate metabolism or mitosis, or prime the cell to respond to another molecular cue.

4.7.2.2 Homophilic Binding: Interactions between Like Molecules

Molecules in immunoglobulin-like families include the vertebrate neural association molecules N-CAM (neural cell adhesion molecule), L1, and N-cadherin, a Ca^{2+}-dependent homophilic molecule. In invertebrates, the *Aplysia* N-CAM-related adhesion molecule plays a role in synaptic plasticity (Mayford et al. 1992).

Grasshopper fasciclins (semaphorins) mediate homophilic adhesion in *Drosophila* (Bastiani et al. 1987; Patel et al. 1987; Harrelson and Goodman 1988; Grenningloh et al. 1990; Hortsch and Goodman 1991; Kolodkin et al. 1992). Connectin, another homophilic molecule, is expressed during *Drosophila* embryogenensis on some neurons and the muscles they innervate (Nose et al. 1992), in longitudinal glia, and in a few neurons that form a scaffold upon which tracts of connectin-expressing axons elongate (Meadows et al. 1994).

Lachesin, a member of the immunoglobulin superfamily, is expressed during neurogenesis in grasshoppers (Karlstrom et al. 1993). It appears at the onset of neurogenesis in neuroblasts, ganglion mother cells and neurons and later becomes restricted to a subset of neurons. Lachesin is closely related to the *Drosophila* amalgam molecule.

Antibodies to homophilic molecules can substitute for the homophilic interactions and activate second messenger systems (Schachner 1991). For instance, antibodies to L1 or N-CAM activate the PIP_2/IP_3 second messenger system, reducing intracellular pH and opening Ca^{2+} channels. The same alterations of IP_3, Ca^{2+} and pH are seen when a cell suspension is added to a monolayer of rat pheochromocytoma cells, and can be blocked by the phorbal ester TPA, which suggests involvement of G protein and protein kinase C (Schuch et al. 1989; Schachner 1991). Doherty et al. (1991) report that effects on the cytoskeleton mediated by N-CAM and L-cadherin binding follow G protein-activation of Ca^{2+} channels. The possible relevance of binding and subsequent alteration in Ca^{2+} level are discussed in Section 4.4.9.

4.7.2.3 Heterophilic Binding: Interactions between Unlike Molecules

The family of molecules called integrins are integral membrane proteins that are involved in a wide range of binding interactions (Venstrom and Reichardt 1993). They have no structural similarity to the molecules that exhibit homophilic interactions nor the extracellular matrix molecules. Integrins mediate adhesion of cells to extracellular matrix molecules such as fibronectin, laminins, tenascin, vitronectin, thrombospondins, proteoglycans and various collagens (Venstrom and Reichardt 1993). Some integrins interact with only one class of matrix molecules but others interact with several. Some interactions involve linkages with the cytoskeleton.

In the leech, at least two and probably several substrate recognition systems are present on neuronal membranes. Binding of cultured leech neurons to Con A (which supports growth although it is not a natural substrate molecule) can be blocked by a specific sugar that does not intefere with binding to leech laminin (Chiquet and Nicholls 1987). The extracellular matrix of the leech contains tenascin and leech laminin, both of which support attachment and sprouting (Masuda-Nakagawa et al. 1988; Masuda-Nakagawa and Wiedemann 1992). In chick development, tenascin (also called cytotactin) is important in preforming

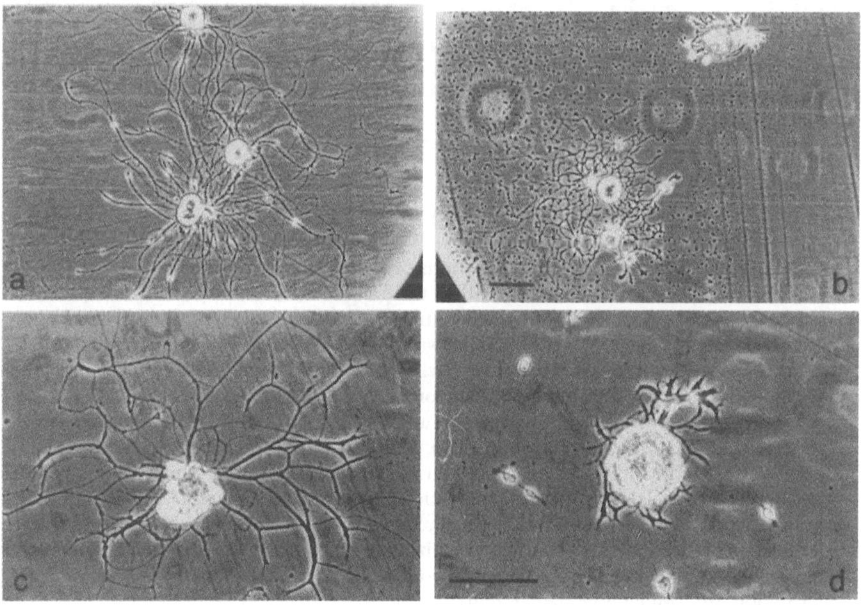

Fig. 4.18. Inhibition of neurite outgrowth by antiserum against a polypeptide (5220 kDa). Retzius neurons were grown for 5 days on dishes coated with ganglion capsule EDTA extract (**a** and **b**) or for 2 days on an affinity-purified laminin-like molecule (**c** and **d**) in the presence of preimmune serum (**a** and **c**) or antiserum against the 5220 kDa subunit of laminin-like molecules (**b** and **d**). Scale *bars* = 100 μm. (Representative cells are shown; quantitative data are given in Chiquet et al. 1988)

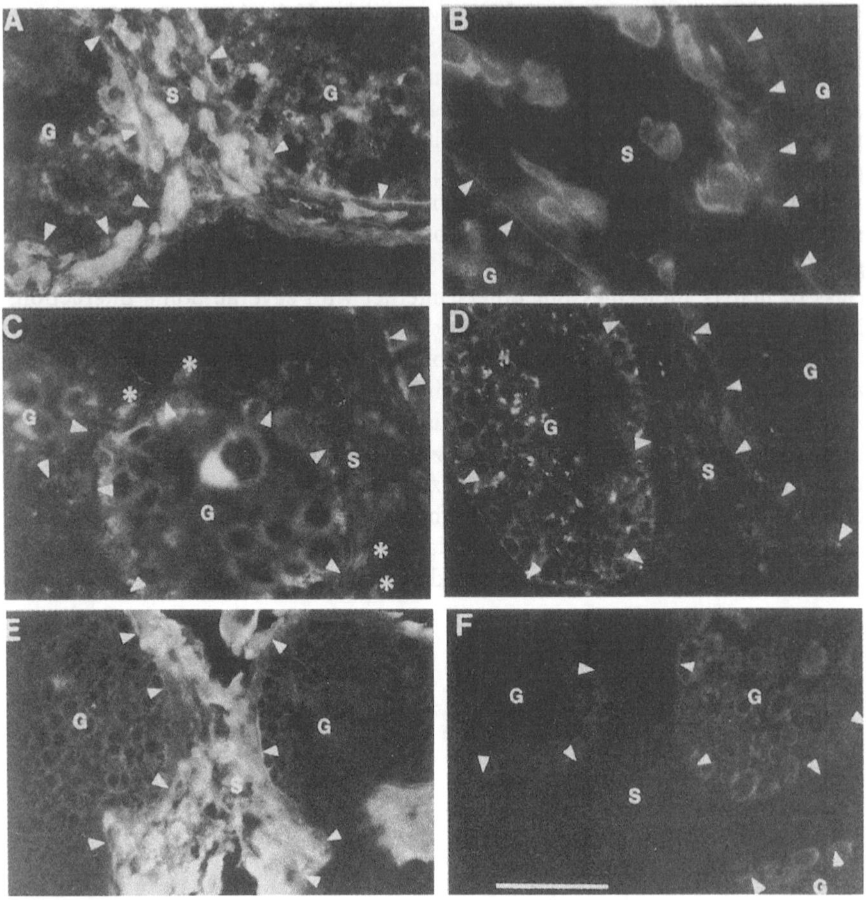

Fig. 4.19. Immunofluorescence of frozen sections of *Helisoma* central ganglionic rings. **A** Anti-laminin antibody: extracellular matrix within the ganglionic sheaths (*S*) and basal laminae (*arrowheads*) surrounding each ganglionic neuropil (*G*) were distinctly labeled. In addition, nonneuronal cells within the ganglionic sheaths exhibited cell surface staining. **B** Higher magnification of labeling of sheath-cell surface with anti-laminin antibody. Labeled basement membrane separated the labeled sheath cells from adjacent neuropil. **C** Anti-B chain antibody: surface labeling of sheath cells (*asterisks*) surrounding the neuropil was consistently seen with this antibody, although the labeling was much weaker than that seen with antilaminin antibody. **D** Control rabbit anti-rat IgG: only autofluorescence is apparent. **E** Anti- ~ 300 kDa antibody: extracellular matrix within the ganglionic sheaths and basal laminae surrounding the ganglionic neuropil were intensely labeled. Surfaces of the nonneuronal sheath cells were labeled but this labeling was difficult to visualize due to the intense fluorescence of adjacent matrix. **F** Nonimmune control rabbit IgG. **E** and **F** were a separate experiment from **A-D**. *Bar* = approx. 300 μm. (Miller and Hadley 1991)

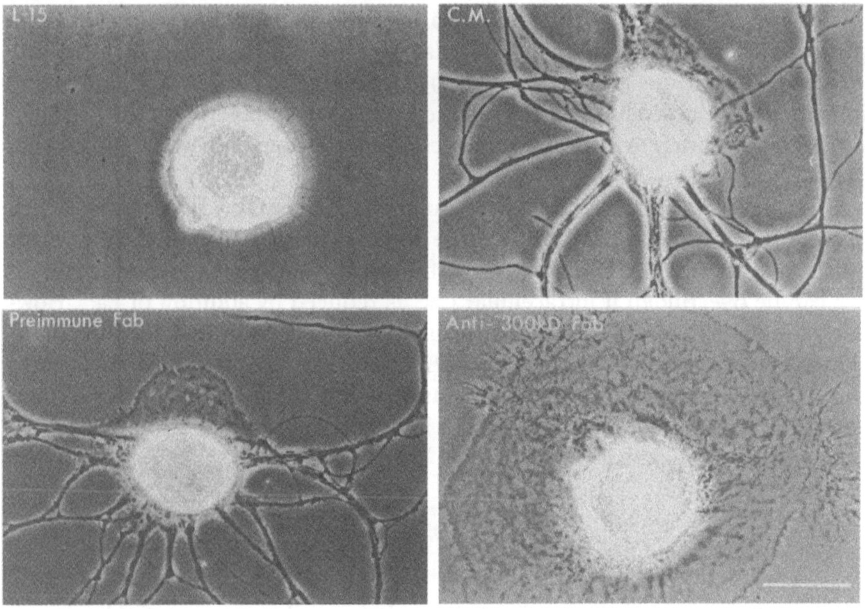

Fig. 4.20. Antibodies against the extracellular matrix protein of *Helisoma* that is recognized by anti-laminin antibodies (Anti 300 kDa Fab fragments) inhibit initial outgrowth of neuron B5 plated in conditioned medium (*C.M.*). All four neurons shown were cultured for 2 days under the indicated conditions. Neurons plated in defined medium (*L-15*) without conditioning do not extend neurites, whereas those plated in *C.M.* or in conditioned medium preconditioned with 100 µg/ml preimmune Fabs (*Preimmune Fab*) both generate extensive neurites. Neurons plated in conditioned medium in the presence of 100 µg/ml *anti- ~ 300 kDa Fabs* never extended normal neurites, but usually had extensive membrane veils around them, like fired eggs. *Bar* = 50 µm. (Hadley, unpubl.)

neural projections; present in leech neuron culture, it allows outgrowth but prevents branching (Wehrle and Chiquet 1990). Leech neurons exhibit different growth patterns on tenascin, laminin, and Con A (Fig. 4.18; Chiquet et al. 1988; Masuda-Nakagawa and Nicholls 1991).

Insects possess a large number of extracellular matrix molecules, including collagens, laminins, fasciclins, tenascin, semaphorin II and connectin. Both culture medium and embryos have yielded laminin-like molecules (Fessler et al. 1987; Montell and Goodman 1988). The expression of fasciclin III on both the growth cone and restricted regions of the target muscle precedes formation of neuromuscular junctions in *Drosophila* (Halpern et al. 1991). The contact also elicits stereotypic sprouting patterns, some of which prefigure the eventual synaptic morphology (Chap. 5).

Molecules similar to mammalian tenascin have been found in the developing olfactory system of *Manduca sexta* (Krull et al. 1994). The tenascin-like molecules are localized on glia, where they form boundaries that shape the developing glomeruli (Tolbert and Oland 1990).

Neural adhesion in *Drosophila* also involves neurotactin, which has a repeated LRE tripeptide sequence similar to that which underlies S-laminin adhesive interactions (Barthalay et al. 1990). Collagen IV is a major component in *Drosophila* extracellular matrix, and ten genes were found when the *Drosophila* genomic library was probed with chicken collagen 1 (Fessler and Fessler 1989).

In the snail *Helisoma*, the extracellular matrix contains fibronectin, which promotes attachment but little neurite outgrowth (Mattson and Kater 1988; Mattson et al. 1988b). The inability of a fibronectin substrate to support sprouting can be overridden if intracellular Ca^{2+} is reduced by addition of La^{3+}. However, the pattern of outgrowth that results is distinct from that seen in conditioned medium (Fig. 4.13). The CNS of *Helisoma* also contains a protein related to vertebrate laminin (Fig. 4.19; Miller and Hadley 1991). Growth cones of neurons growing on *Helisoma* extracellular matrix collapse when antibodies to the native protein are added (Fig. 4.20).

4.7.2.4 Identification of Glycoproteins

As indicated above, glycoproteins (proteoglycans) play important roles in cell adhesion and recognition in the nervous sytem (Jessell et al. 1990). Plant lectins and antibodies offer two experimental approaches to detecting and characterizing glycoproteins. Although many glycoproteins are currently described only in connection with their role in development, some glycoproteins may be reexpressed in regenerating nervous systems.

Different lectins bind to different sugars. For instance, Con A is a lectin that recognizes the terminal sugars α-D-glucose and α-D-mannose. This binding may substitute for interactions with natural substrate molecules and account for the effectiveness of Con A as a neuron culture substrate. Wheat germ agglutinin (WGA) recognizes both N-acetyl-D-glucosamine and N-acetylneuraminic acid, and peanut agglutinin (PNA) recognizes β-D-galactose. Although these markers respond to a definable entity, they may fail to distinguish between two different glycoproteins with the same terminus, whereas natural receptors may be more discriminating due to binding with other sites.

Lectins show that a variety of molecules on the surfaces of neurons are potentially involved in adhesion and target recognition, because many surfaces bind more than one lectin. For instance, embryonic locust neurites bind Con A, WGA and PNA, as well as being reactive to laminin (Vanhems and Delbos 1989). The specificities in neuromuscular innervation patterns in arthropods have been revealed by glycoprotein binding of lectins (Denburg et al. 1983a,b; Denburg et al. 1987).

In *Aplysia* neurons, several populations of glycoproteins have been separated, including one associated with the nuclear membrane, and others that are transported down the axon (Gabel et al. 1989). The specificity of the binding of the different neuronal glycoproteins to target cells was investigated by extracting them and coupling them to fluorescent spheres. Microspheres loaded with

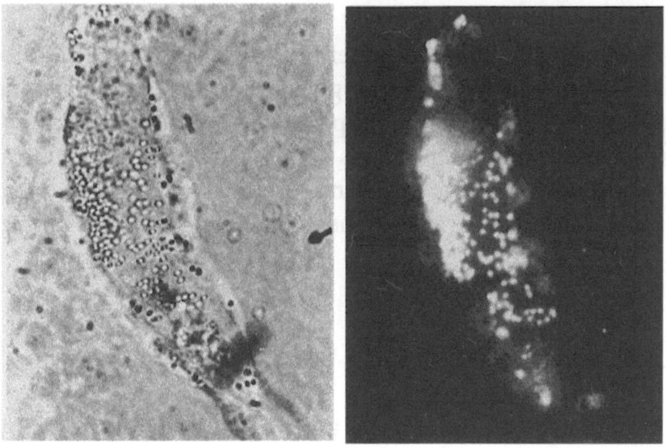

A B

Fig. 4.21. Binding of glycoprotein wheat germ agglutinin (GPwga) to isolated auricle muscle cells. GPwga, isolated from the membranes of abdominal ganglion neurons, and glycopeptides obtained from ovalbumin (GPov), were coupled separately to 0.9 μm fluorescent beads. Auricle muscle cells were isolated by treatment of auricle muscles with collagenase and were grown in vitro for 24 h. The cells were then fixed with 1% paraformaldehyde for 1 h. After thorough washing with seawater, equal amounts (based on neutral hexose) of GPwga- and the ovalbumin-coupled beads were added to the cells. The muscles were washed 2h later with seawater and examined. **A** Light micrograph of an auricle cell that had been exposed to the GPwga-coupled beads. The *black beads* can be seen bound to the surface. **B** The same cell viewed with fluorescence optics. There was no significant binding of GPov to the muscle cells. (The binding of the conjugated beads is quantitated in Ambron et al. 1989)

glycoproteins known to be transported down to growth cones of left upper quadrant cells were found to bind to the target tissue, auricle muscle cells (Fig. 4.21; Ambron et al. 1989).

Glycoproteins can also be detected with antibodies. It is usually necessary to generate antibodies against the native protein because vertebrate antibodies may react poorly, or not at all, with invertebrate forms of the molecules (Chiquet et al. 1988; Mattson and Kater 1988; Fessler and Fessler 1989). Crude neural or connective tissue capsule homogenates provide material to generate a wide selection of monoclonals.

A monoclonal antibody (DSS-3) that binds to all cells in the neuronal lineage of the cockroach nervous system is present during the period of axonal elongation, but antigenicity is then lost in all but a subset of neurons Denburg and Norbeck 1989; Denburg et al. 1989). However, the antibody reappears in regenerating sensory neurons, motoneurons and interneurons following axonal injury, and antigenicity is retained in motoneurons until they have eliminated inappropriate connections (Denburg 1989).

The role of the extracellular matrix in guiding *Drosophila* axons in embryonic and metamorphic stages is reviewed by Fessler and Fessler (1989) and Truman (1990). An antibody to a glycoprotein, 66B2, exhibits strong antigenicity in the extracellular matrix only during these stages, coincident with axonogenesis and synaptogenesis (Go and Hotta 1992).

Glia in the nervous system of the insect *Tenebrio molitor* show a temporal pattern in expression of the glycolipid-derived glucuronic acid-containing epitope, CAF-1, in larval and pupal stages (Breidbach et al. 1992). The same epitope is detected following lesions to leg nerves or the central neuropil; it is reexpressed within 6 h of lesioning in restricted regions, including the termination areas of the regenerating sensory neurons and the ventral association areas.

4.7.2.5 Antibodies, Blockers and Mutations Disrupt Pathfinding

Manipulation of developing and regenerating systems by the blocking of epitopes with plant lectins and antibodies has revealed the functions of particular molecules. If a substrate molecule is demonstrated immunohistochemically to be present in the environment of growing neurites, its relevance to the process of development or regeneration can be tested by adding antibodies (Chiquet et

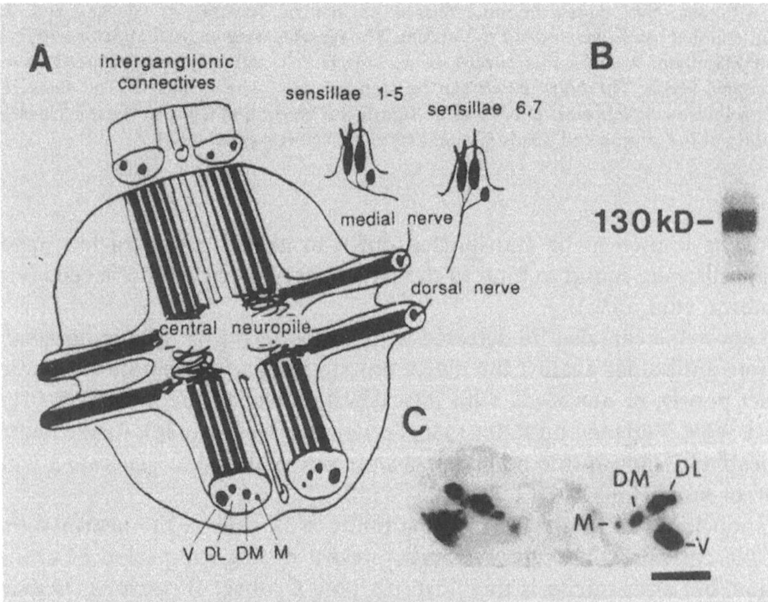

Fig. 4.22. Sensory afferents in midbody segment of the ED12 germinal plate. **A** Midbody sensillae contain the cell bodies of three types of sensory afferents: photodetector without cilia, uniciliated movement detectors, and a multiciliated type of undetermined function. Their axons project as tightly fasciculated tracts into the segmental ganglion. In the central neuropile, afferents spread into a diffuse tree before reassembling into four ipsilateral neuropile tracts: the ventral (*V*), dorsomedial (*DM*), dorsolateral (*DL*), and thin medial (*M*). Sensory afferents project out of the neuropile into the interganglionic connective. **B** A Western blot of CNS extract was stained with Lan3–2 Fab fragments, which recognize the 130 kDa sensory afferent protein. **C** A cross section of an ED12 embryonic ganglion demonstrates the four central neuropile tracts stained by MAb Lan3–2. *Bar* = 10 μm. (Zipser et al. 1989)

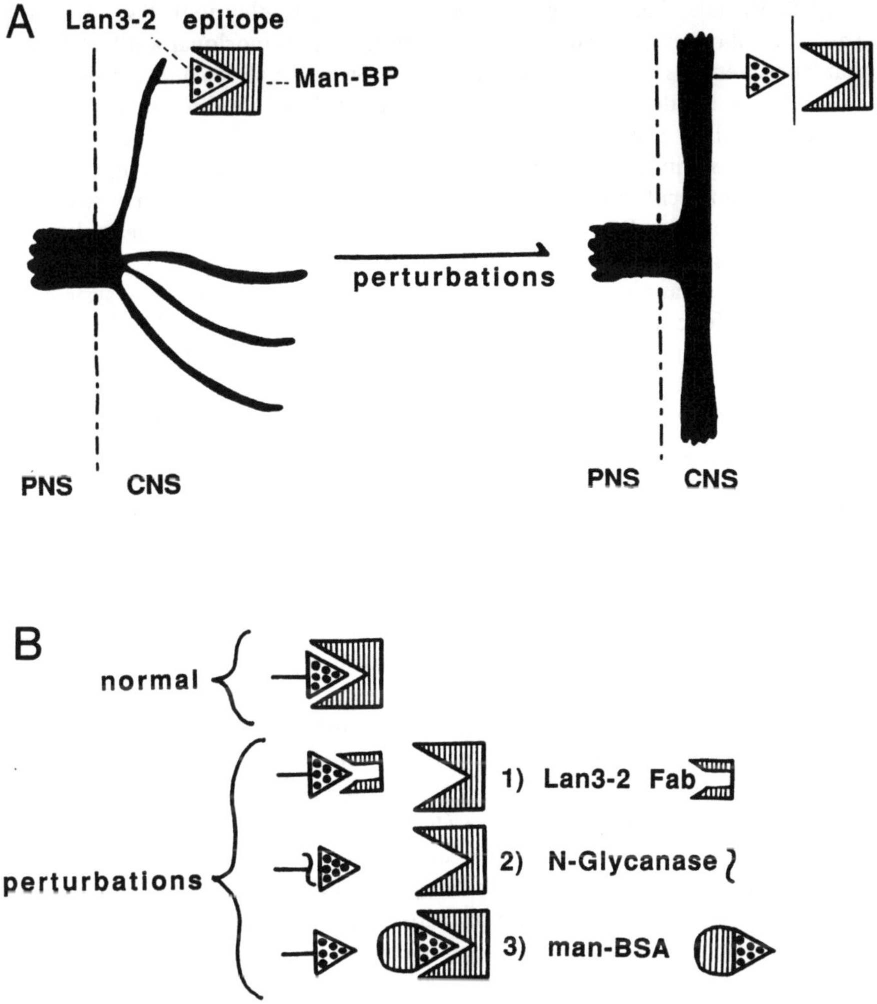

Fig. 4.23. Hypothesis of carbohydrate recognition mediating the defasciculation of sensory afferents into the synaptic neuropil. **A** During development, sensory afferent axons grow as a fasciculated bundle through a peripheral nerve (*PNS*) into a central ganglion (*CNS*), where they defasciculate, projecting as single axons through the synaptic neuropil. Even though the mannose-containing *Lan3-2* epitope is evenly distributed on the sensory afferent surface, the defasciculation event is restricted to the synaptic neuropil. Our experimental evidence is consistent with a mannose-binding protein (*Man-BP*) being present in the synaptic neuropil. Recognition/adhesion between the *Lan3-2* epitope of sensory afferents and the *Man-BP* in the synaptic neuropil results in the region-specific defasciculation. Perturbations of this mannose-specific recognition inhibit the defasciculation, and, instead, sensory afferents continue to grow through the CNS as a bifurcated fasciculated tract, appearing as a T. **B** The normal recognition between the *Lan3-2* epitope and the *Man-BP* is inhibited in the presence of (*1*) *Lan3-2 Fab* fragments; (*2*) *N-glycanase*, a glycosidase; and (*3*) mannose-BSA (*man-BSA*), a neoglycoprotein. The *Lan3-2 Fab* fragments bind to and mask the mannose-containing *Lan3-2* epitope, *N-glycanase* cleaves N-linked carbohydrate domains removing the *Lan3-2* epitope, and *man-BSA* competes with the *Lan3-2* epitope for the mannose-binding protein. (Zipser and Cole 1991)

al. 1988; Masuda-Nakagawa et al. 1990; Miller and Hadley 1991; Figs. 4.18 and 4.20). Grasshopper embryonic tissue has yielded a monoclonal antibody, 2B2, that labels neurons during axonal elongation (Seaver et al. 1991). Addition of the antibodies to the developing system interferes with pathfinding for the growth cones that follow preexisting axonal pathways, but not for the pioneers that depend upon nonneuronal substrates. Similarly, a developmental stage-specific cell surface molecule that is also found in the extracellular matrix which is detected by antibody DSS-8, functions in pioneer axon growth in the cockroach and, later, in selective fasciculation (Wang et al. 1992).

Antibodies have been added to the developing leech nervous system to determine if function is perturbed (McRorie and Zipser 1988). Leech sensory neurons share a 130 kD glycoprotein that has a mannose epitope asparagine-linked to the protein (Fig. 4.22). The Lan3-2 antigen of this glycoprotein is present on the surface of most sensory cell somata, neurites and axons (Zipser et al. 1989). Sensory neurons travel together in (fasciculated) bundles from the periphery to the CNS (Zipser and Cole 1991). There, adhesion gives way to defasciculation, as sensory fibers fan out in diffuse projections in the vicinity of their target cells before aggregating to form central tracts (Zipser and Cole 1991). The defasciculation can be blocked by the Lan3-2 antibody and other procedures that remove the mannose epitope (Fig. 4.23). This suggests that, upon entering the ganglion, neurites normally encounter a mannose-binding protein which provides sufficient attraction to overcome the interactions responsible for fasciculation.

Within the ganglion, leech sensory axons refasciculate into four tracts to project to other levels within the CNS. The antibody Lan4-2 discriminates one of the fascicles that form within the CNS, suggesting that a hierarchy of guidance cues mediate specific pathways (Johansen et al. 1992). At least some of these tracts have been established by central neurons before the arrival of sensory cells (Jellies et al. 1994). Such behavior must involve discrimination of positional cues by sensory subgroups (Zipser and Cole 1991). Sensory modality has been correlated with the possession of different carbohydrate epitopes (Zipser et al. 1994).

The sex afferents are a distinct class of sensory cells associated with the sex organs. These afferents normally enter only the fifth and sixth body ganglia, where they form a fifth tract. Transplantation experiments reveal that the information necessary for these afferents to form this tract is also present in other body ganglia (Passani et al. 1991).

The generation of sensory neurons throughout life in leeches is correlated with the retention of regional cues which govern fasciculation and defasciculation. The presence of such cues accounts for the successful regeneration of leech sensory afferents (Peinado et al. 1987a, b; Holder and Clark 1988). Crushing a posterior nerve root provides regenerating axons of peripherally-located sensory neurons with the choice of growing within two separate Lan2-3-positive tracts or growing elsewhere in the nerve. The axons do not discriminate between their "correct" tract and the other tract, but do choose one of the paths that could be identified by the residual Lan2-3 immunoreactivity (Fig. 4.24).

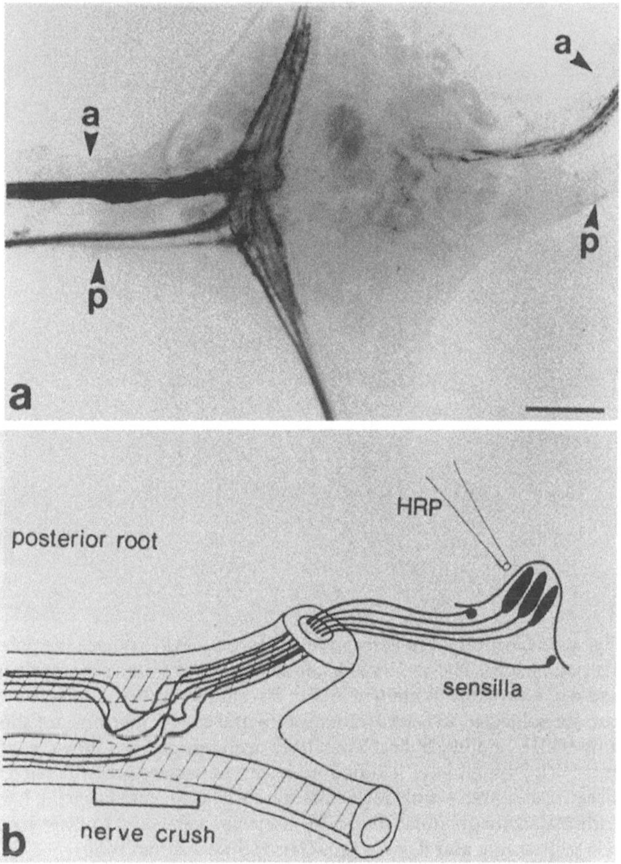

Fig. 4.24. a Micrograph of a whole-mount ganglion preparation showing the normal projection of peripheral sensory axons (*left side*), and regenerating axons filled with HRP 10 days after a crush (*right side*). Two tracts are visible in the left posterior root. As is often the case, axons in the posterior root nerve have not regenerated. *a* Anterior roots; *p* posterior roots. *Bar* = 100 μm. **b** A posterior root nerve in which a choice between two *Lan3-2* positive tracts is made by regenerating axons. The leeches used in this study were approximately 1 year old and weighed between 0.5 and 1.5 g. Animals were anaesthetized in 0.15 mM chlorobutanol (Sigma) and a small incision made on the ventral body wall over a midbody ganglion. The nerve cord was pulled out of the body enough to expose the root nerves which then were crushed with a fine pair of forceps at a distance of about 200–300 μm from the edge of the ganglion. Animals were kept in water at 18°C. Ten days after crushing the nerves, leeches were anesthetized and placed ventral side up on a dry surface. The cuticle was perforated with a sharp minuten pin at several places on the ventral aspect of the middle annulus in the segment to be labeled. A solution of 30% HRP (Boehringer, Grade II) and 3% lysolecithin (Sigma) in distilled H_2O was then applied to the skin and uptake allowed to take place for 2 h at 2–4°C. Animals were then returned to seawater and kept overnight at 18°C in order for the tracer to diffuse inside the axons. The following day ganglia were dissected out, fixed in 4% paraformaldehyde in 0.1 M phosphate buffer (pH 7.4), and processed for HRP with cobalt intensification. (Peinado et al. 1987a)

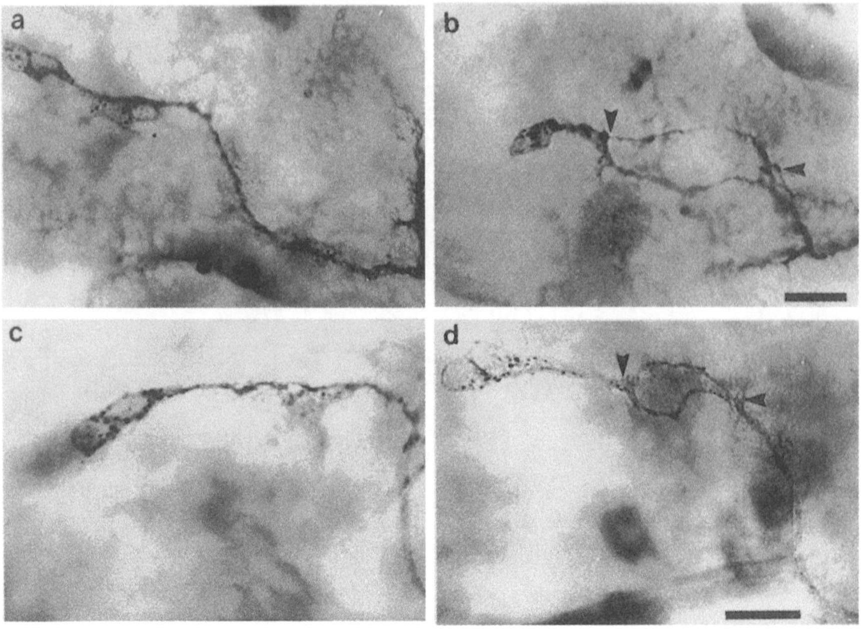

Fig. 4.25. Chromophore-assisted inactivation (CALI) causes the defasciculation of grasshopper Ti1 sister axons. The *arrows* indicate the points of axons separating in CALI treated embryos (**b** and **d**); these axons remain as single fascicles in embryos injected with dye-labeled antibodies but not subjected to laser irradiation (**a** and **c**). Ti1 neurons are shown from the metathoracic limb buds of sibling embryos that have been injected as follows: **a** with Malachite green-labeled anti-HRP without laser irradiation; ***b*** with Malachite green-labeled anti-HRP and subjected to laser irradiation; **c** with Malachite green-labeled anti-fasciclin I without laser irradiation; **d** with Malachite green-labelled anti-fasciclin I subjected to laser irradiation. Scale *bar* in **a** and **b** = 30 μm; in **c and d**, = 40 μm. (Jay and Keshishian 1990)

Another technique for deducing function by disrupting normal processes utilizes laser inactivation to selectively destroy bound protein that has been labeled with an antibody-bearing chromophore (Jay and Keshishian 1990). Heat released from the laser-excited dye is transmitted to the antibody-bound transmembrane molecule, resulting in its denaturation. Fasciculation of pioneer neurons in the grasshopper embryo is mediated by fasciclin I, and when it is denatured by chromophore-assisted laser inactivation, defasciculation results (Fig. 4.25).

Pathfinding can also be altered by altering the expression of a receptor molecule. In *Caenorhabditis elegans*, *unc-5* gene is responsible for a cell surface receptor responsive to dorsal pathway choices labelled by expression of *unc-6* (Hedgecock et al. 1990). When *unc-5* is ectopically expressed in touch cells of transgenic animals, their axons are steered dorsally, rather than in the normal longitudinal or ventral trajectories (Hamelin et al. 1993). Many studies of mutants in *C. elegans* provide additional information on axon guidance (e.g., Hedgecock et al. 1987; Hekimi and Kershaw 1993).

4.7.2.6 Cues Expressed in Gradients

The molecules responsible for substrate binding provide a permissive pathway but do not appear to attract growing neurites on the basis of a graded distribution. This has led to a search for new categories of molecules. Among vertebrate preparations, the netrins influence the growth of neurons in the spinal cord through gradients (Baier and Bonhoeffer 1994). The genes for the netrins bear a remarkable resemblance to the *unc-6* gene of *Caenorhabditis elegans* (Hedgecock et al. 1990), which controls interactions among growing axons that orient them within the animal.

The pioneer neuron Ti1 of insect (orthopteran) limbs grows centrally to establish the leg nerve pathway through aneural territory (Bate 1976; Keshishian 1980). The morphology of the pioneer growth cones of this cell suggest that they respond to a gradient of cues within the leg segments (Caudy and Bentley 1986). In contrast to the homogeneous distribution of many substrate binding molecules, gradients of three molecular species appear prior to and during the outgrowth of pioneer axon Ti1 along the axis of the cockroach leg (Norbeck et al. 1992). These molecules, which were identified by hybridoma techniques applied to embryonic tissue, are good candidates for pioneer axon guidance cues. Monoclonal antibodies to these three basal lamina components bind to the basal lamina/extracellular matrix in a graded pattern. One binds the distal part of the leg more intensely, whereas the other two label the proximal part of the leg more intensely.

4.8 Conditioned Medium and Growth Promotors

4.8.1 Growth Requirements in Culture

Culture of neurons in vitro has helped to define growth requirements (see Beadle et al. 1988). Short-term survival of most invertebrate neurons requires a balanced saline, which can be supplemented with sugars and amimo acids. In a defined medium such as Leibowitz L15, most neurons isolated without a length of axon round up and will not sprout. The nonsprouting neurons retain fairly normal electrical properties. Like vertebrate neurons, some invertebrate neurons require trophic factors for survival.

Growth in culture typically requires both a suitable substrate and promotors. Developing and regenerating neurons must attach to some surface, which can even be a fluid-air interface, before extending neurites. Stomatogastric ganglion neurons isolated from crabs and lobsters attach both to sterile plastic dishes and poly-L-lysine coated dishes (Graf and Cooke 1990), and outgrowth is seen only from sites of attachment (Fig. 4.26). Although attachment is important, an overly attractive substrate can block responses to other signals. Neu-

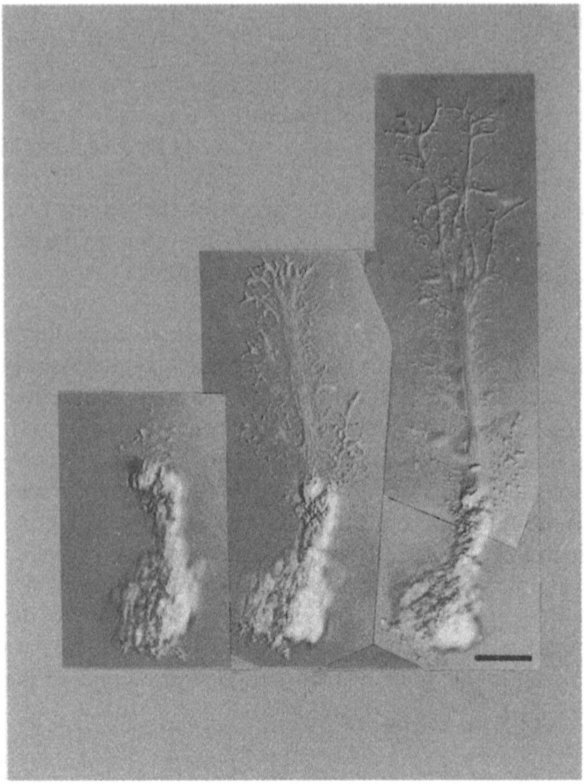

Fig. 4.26. Hoffman modulation contrast images of directed outgrowth from a *Cardisoma carnifex* stomatogastric neuron at different culture ages (*left* to *right*; day 1, day 2, day 3) in static-defined medium. At day 1, filopodia are seen at the peripheral margin of the small veil. By day 2, there has been extensive directed outgrowth from day 1 filopodia. Further extension of several processes occurred by day 3. Note the fibrillar appearance of the stabilized major neurite. Scale *bar* = 50 μm. (Graf and Cooke 1990)

rons growing on highly adhesive substrates have very flattened growth cones and may be incapable of responding to additional cues in the environment (Bray 1991).

There is increasing evidence for homologies between the molecules that control development and growth in invertebrates and vertebrates (Muskavitch and Hoffman 1990). For instance, epidermal growth factor (EGF) stimulates mitosis in cells derived from brains of adult mice, producing large numbers of neurons and astrocytes in culture (Reynolds and Weiss 1992). EGF has homologs in flies *(Drosophila)*, nematodes (*Caenorhabditis*) and sea urchins (*Strongylocentrotus*). Striking similarities in the functions controlled by these homologs to EGF have emerged. Growth factors of invertebrates have in some cases been revealed by mutants, which have facilitated analysis of relationships between molecular structure and function.

4.8.1.1 Growth Factors in Gastropod Conditioned Medium

Growth factors in blood and brain-conditioned media were discovered because they stimulated the growth of cultured neurons. The *Helisoma* neuron culture medium is produced by incubating isolated brains in osmotically adjusted Leibowitz-15 medium (2 brains/ml) for 72 h, and adding the medium so obtained to polylysine-coated culture dishes for 24 h (Wong et al. 1981). Condi-

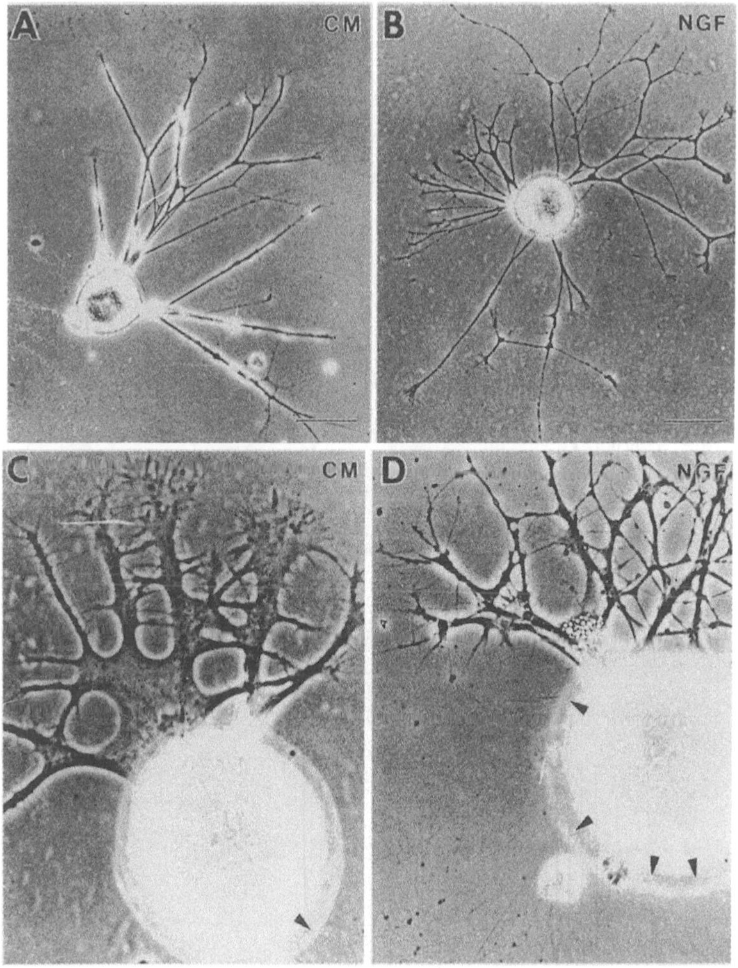

Fig. 4.27. Qualitative nature of the response to NGF by pedal A cluster neurons of *Lymnaea*. The response of motoneurons and interneurons to NGF (**B**) in most respects resembles the sprouting activity observed in **A**, conditioned medium (*CM*). Close examination of neurons in NGF-supplemented medium, however, reveals a striking difference in the nature of the outgrowth compared to *CM*. In the presence of NGF (**D**), primary neurites and, especially, cell bodies elaborated many fine extensions (*arrowheads*) that were much less frequently observed in *CM* (**C**). Scale *bars* in **A** and **B** = 100 μm; in **C** and **D** = 50 μm. (Ridgway et al. 1991)

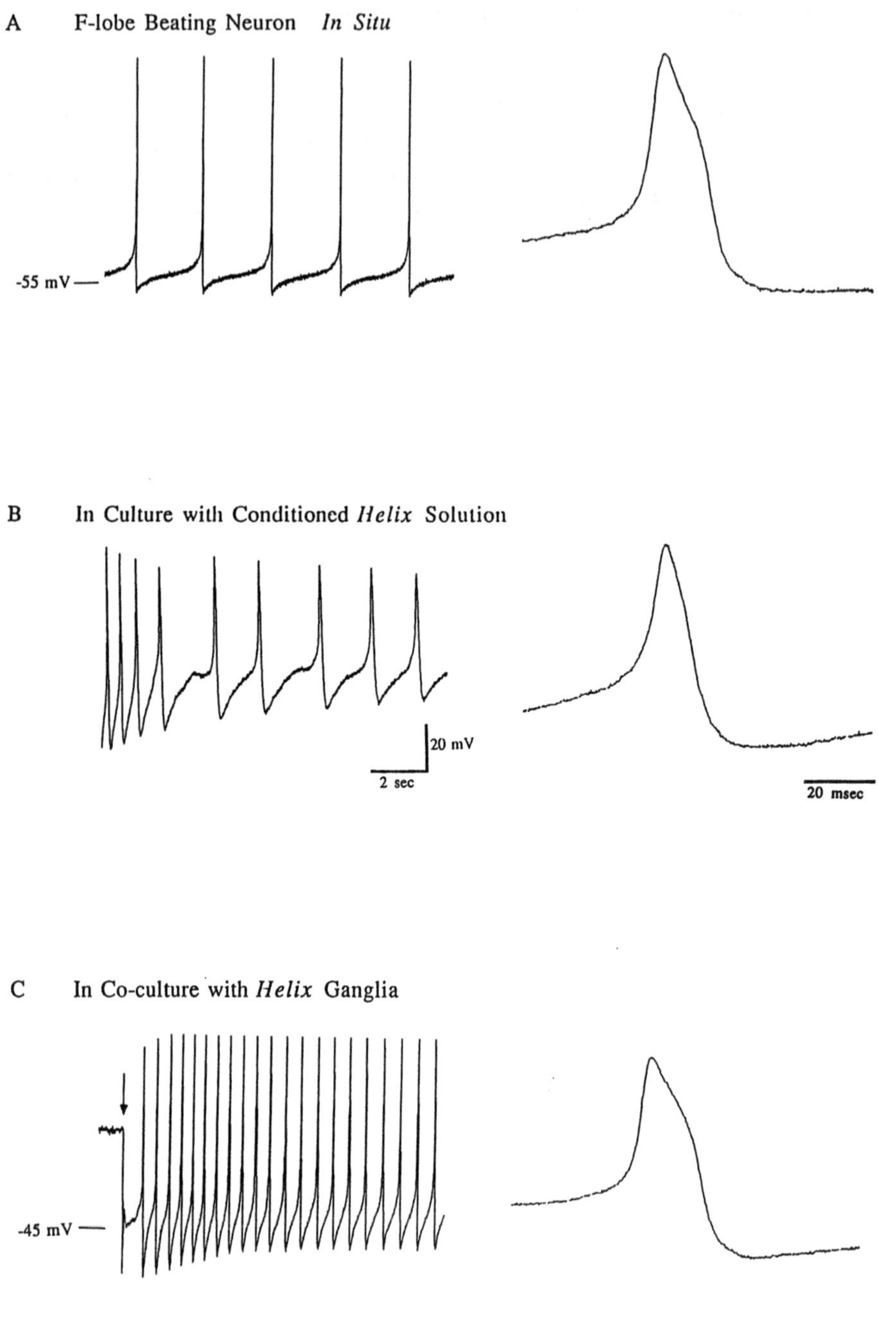
A
F-lobe Beating Neuron *In Situ*
-55 mV
B
In Culture with Conditioned *Helix* Solution
20 mV
2 sec
20 msec
C
In Co-culture with *Helix* Ganglia
-45 mV

tioning factors are deposited as a substrate during this period; the conditioning solution is then removed, the dishes washed, and isolated neurons in unconditioned medium are added.

Helisoma conditioned medium contains multiple growth-promoting factors, some soluble and others capable of binding tightly to the substratum (Barker et al. 1982; Bulloch and Hauser 1990). Release of growth-promoting factors from *Helisoma* ganglia depends upon electrical activity (Berdan and Ridgway 1992). Interfering with the spontaneous electrical activity by adding either $CoCl_2$ or tetrodotoxin (TTX) significantly reduces the number of neurons that sprout, compared with the medium obtained under control conditions.
Although *Helisoma* conditioned medium supports the sprouting of all classes of neurons, different classes of neurons have different requirements. Sprouting of motoneurons and interneurons is reduced by antibodies against NGF (Fig. 4.27; Ridgway et al. 1991). In the vertebrate system, it is the sensory neurons that exhibit up-regulation of NGF receptors and increased expression of NGF in the denervated target (Mcarow et al. 1994). Whether NGF expression is increased in regenerating snails remains to be determined, but the evidence for an NGF-like factor supports the impression that several growth factors are present in conditioned medium, already known to include laminin-like molecules. Antibodies against *Helisoma* laminin block outgrowth of buccal ganglion neurons B5 in conditioned medium (Miller and Hadley 1991; Fig. 4.20). When Fab antibody fragments generated against the approximately 300-kDa laminin-like component of *Helisoma* conditioned medium are added to neurons growing on conditioned medium, growth cones stop growing or collapse and retract.

Neurons of the snail *Helix* cultured in conditioned medium prepared according to the technique developed by Wong et al. (1981) have electrical properties inferior to those neurons cocultured with *Helix* ganglia (Tiwari and Woodruff 1992; Fig. 4.28) Presumably, a factor continually released by the ganglia is necessary.

◀ **Fig. 4.28.** Effect of "conditioned *Helix* Ringer's solution" on the electrical activity of a right parietal beating neuron in culture. **A** Typical beating behavior of a right parietal cell in an intact ganglion. In order to show the shape of an action potential, an expanded view of part of the record appears on the *right*. **B** A record from a neuron in culture for 2 days with "conditioned *Helix* Ringer's solution". Nine action potentials were triggered by injecting 0.2 nA of positive current through the recording pipette. The action potential on the *right* is the third action potential from this series. The resting potential was uncertain in this trace because a bridge circuit was not used. **C** A record from a neuron after 2 days in culture with "conditioned *Helix* Ringer's solution" plus *Helix* ganglia (coculture). Spontaneous action potentials occurred immediately after penetration of the cell with the microelectrode (*arrow*). The shape of one of these spontaneous action potentials is shown on the *right*. The voltage and time scales in **B** apply to **A** and **C** as well. (Tiwari and Woodruff 1992)

The hormones somatostatin and calcitonin were found to support the production of neurites from isolated neurons of the snail *Physella* (Grimm-Jorgensen 1987) and somatostatin similarly promotes neurite outgrowth of *Helisoma* neurons in cultured ganglia (Bulloch 1987). Both hormones act in vertebrates by reducing intracellular free Ca^{2+} (reviewed by Grimm-Jorgensen 1987) and immunoreactivity to both of these hormones has been demonstrated in gastropods (Grimm-Jorgensen 1983; Bulloch 1987).

4.8.1.2 Neurosecretory Products as Growth Factors

If the neurons to be cultured are themselves hormone-releasing, might they be able to satisfy their own requirement for growth factors? The one group of neurons in *Aplysia* known to sprout in defined medium is the peptidergic bag cells (Kaczmerek et al. 1979). Similarly, peptidergic neurons from the eyestalk neurosecretory system of crabs and lobsters grow in defined medium, and the same is true of the peptidergic stomatogastric ganglion neurons (Graf and Cooke 1990). Cultured neurons continue to synthesize peptides after transfer to culture conditions (Hall and Lloyd 1991), and secretory neurons would be likely to continue releasing their products in culture. Although the factors released by such cells may not suffice to stimulate sprouting in all neurons of a given nervous system, they may support a select class of neurons. Similarly, an endocrine organ, the dorsal body, releases a factor in coculture with snail neurons that supports sprouting of neurosecretory neurons but not interneurons or motoneurons of *Helisoma* and *Lymnaea* (Ridgway et al. 1989, 1991).

Addition of the amino acid glutamate to the culture medium promotes the release of several large molecules, particularly non-substrate-absorbed growth-promoting molecules (Bulloch 1985; Bulloch and Hauser 1990). One of the large molecules released in response to glutamate shares antigenic determinants with murine nerve growth factor (NGF) (Ridgway and Bulloch 1991). NGF is a neurotrophic factor for mammalian sensory and autonomic neurons and central cholinergic neurons (Thoenen 1991). Similarly, in the snail *Lymnaea*, it induces sprouting of specific classes of neurons: motoneurons and interneurons but not sensory and neurosecretory neurons (Ridgway et al. 1991).

4.8.1.3 Insulin-Like Molecules Support Sprouting

The insulin family apparently arose early in evolution and has a widespread distribution among vertebrates and invertebrates, where its involvement with development and growth is a consistent feature (Ebberink et al. 1989). Membrane glycoproteins related to the human insulin receptor are found in the developing *Drosophila* brain (Piovant and Lena 1988). The insulin-like neuropeptide released by neuroendocrine cells (the light green cells) in the brains of *Helix* and *Lymnaea* controls growth and has growth-promoting effects on cultured neurons (Geraerts and Joosse 1989; Kits et al. 1990; Gomot et al. 1992).

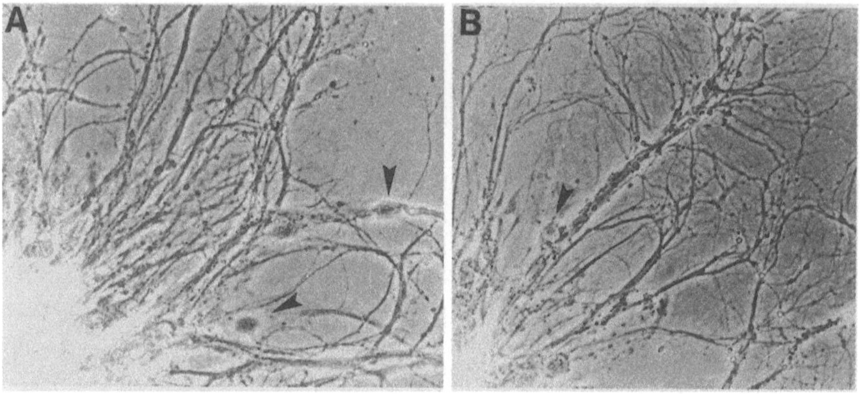

Fig. 4.29. Neurite outgrowth from the cultured locust CNS. Detail of neurite outgrowth in the presence of insulin (**A**) and neuroparsin (**B**). Phase contrast micrographs after 14 days in vitro. The *arrowhead* shows the glial cells beneath the fibers (x600). (Vanhems et al. 1990)

This insulin-like factor, which reacts with an antibody against porcine insulin, is released from cultured *Lymnaea* ganglia and can be extracted from the neurohemal structure formed by the processes of the light green cells in the median lip nerve. Antibodies to synthetic fragments of molluscan insulin-like prohormone (anti-MIPs) have allowed the prohormone to be mapped in the brains of several gastropods, and this has confirmed the neurohemal release sites (Sonetti et al. 1992).

The prothoracicotrophic hormone (bombyxin) identified in the silkworm *Bombyx mori* belongs to the insulin family (Nagasawa et al. 1986). Insulin enhances the growth of locust CNS explants grown in serum-free medium. It interacts synergistically with the molting hormone 20-hydroxyecdysone to promote neurite outgrowth of embryonic locust neurons (Vanhems et al. 1990). The insect neurohormone neuroparsin, which is associated with energy metabolism during migration, similarly promotes neurite outgrowth in synergy with ecdysone, but the morphology of the stimulated growth is distinct for the two factors (Fig. 4.29).

4.8.1.4 Inhibitors of Sprouting

Conditioned medium supplies many of the requirements of neurons in culture, including substrate molecules and growth promotors. It is also likely to contain inhibitors. Blood also contains factors that affect growth, and whether inhibitory or stimulatory influences dominate varies from species to species. A high-molecular-weight factor in *Aplysia* blood promotes neurite outgrowth of *Aplysia* neurons but the presence of *Helisoma* blood inhibits outgrowth of *Helisoma* neurons (Goldberg et al. 1991).

Both neural and nonneural cells release factors that stimulate or inhibit regeneration. For instance, regenerating fish optic nerve and mammalian

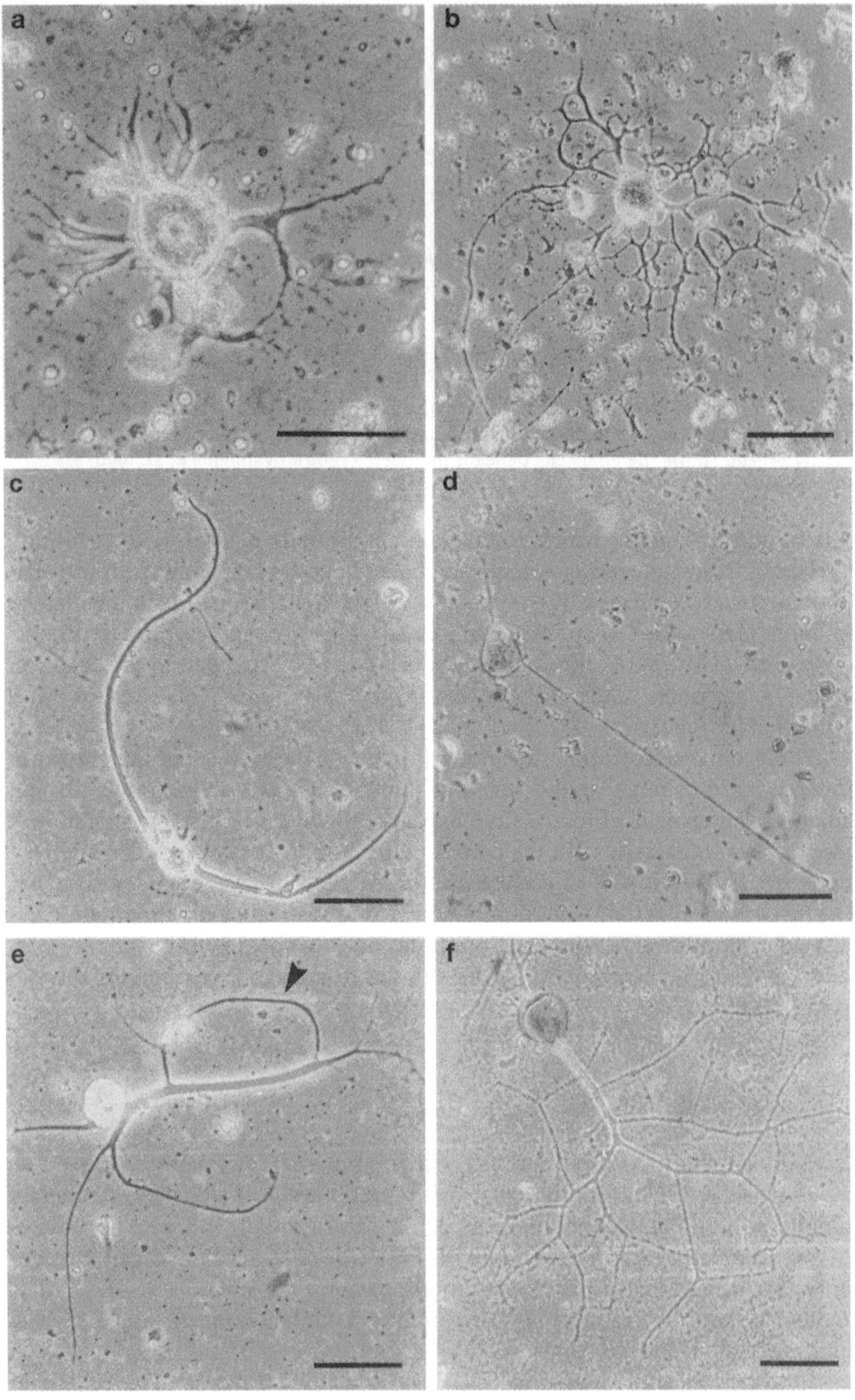
a
b
c
d
e
f

sciatic nerve produce factors that stimulate regeneration in test systems, such as the (normally) unresponsive optic nerve of a mammal (Schwartz 1987). These "growth-associated triggering factors" are produced by nonneural sheath cells rather than the axons. Regeneration of optic tentacles of the snail *Achatina* is correlated with the release of factors into the blood that speed regeneration when transfused into other snails; an inhibitory factor is also present (Sidelnikov and Stepanov 1991, 1992; Moffett et al. 1995).

Blood-borne hormones are known to influence neuron differentiation in insects and the reshaping of dendritic arbors at metamorphosis (Weeks and Truman 1986; Weeks et al. 1992). Neuron death and survival are also under hormonal control. Embryonic insect neural tissue has been amenable to culture (Chen and Levi-Montalcini 1970; Wu 1988; Lees and Beadle 1991; Vanhems et al. 1993) and the culture of *Drosophila* neurons from different parts of the developing nervous system reveal differences in growth cones that may reflect differentiation of axons and dendrites (Kim and Wu 1991). Adult insect neurosecretory cells and others have been successfully cultured (Seshan and Levi-Montalcini 1973; Howes et al. 1991). Fully differentiated thoracic neurons of cockroaches grow on defined medium supplemented with cockroach blood; a layer of blood cells on the bottom of the dish was essential to support sprouting in that system (Fig. 4.30).

Natural growth promotors are present in the medium derived from cultured leech ganglia and in the extracellular matrix. Leech laminin and fibronectin provide sites for substrate attachment and also stimulate outgrowth (Masuda Nakagawa et al. 1988). Con A, which is not a natural component of the extracellular matrix, supports outgrowth if it is in substrate form, but not in solution. Neurons cultured on the border between leech extracellular matrix and Con A project neurites onto each substrate, and the neurite morphologies are modified by the substrate upon which they grow (Grumbacher-Reinert 1989; Fig. 4.5).

In the search for promotors and inhibitors of growth in the culture medium, the observations on planarian and nemertean regeneration should be kept in mind. These suggest that an extract of the tissue added to the culture medium is able to inhibit regeneration of a given tissue (Chap. 2).

◀ **Fig. 4.30.** Phase-contrast light micrographs showing outgrowths from neurons isolated from thoracic ganglia and maintained in vitro on a substratum derived from insect blood. **A** A single cell showing multiple process outgrowth after 5 days in culture. **B** A neuron after 16 days in culture surrounded by a network of processes. **C** An isolated neuron showing bipolar growth after 8 days in culture. This neuron has processes that are characteristically curved and have very few branches. **D** A neuron from a 12-day-old culture showing a single main process that is both straight and unbranched. **E** Branching pattern of an isolated neuron from a 10-day-old culture showing curving branches some of which turn back towards the main trunk (*arrowhead*). **F** Branching pattern of an isolated neuron after 25 days in culture. Here, branches have radiated from the main neurite, and further secondary branches have developed to form a network on one side of the cell (in contrast to the network surrounding the cell in Fig. 3B, where the branching pattern develops from a number of processes). Scale *bars* = 100 μm. (Howes et al. 1991)

4.8.2 Culture Influences on Branching Pattern

Invertebrate neurons have characteristic branching patterns in situ, and often the growth cones and branching pattern they produce in vitro allow them to be distinguished from one another (Haydon et al. 1985; Howes et al. 1991). In situ, cues in the surroundings and the competition with other neurons for target-derived factors determine the fate of individual branches of a neuron's distinctive morphology (Purves et al. 1988; Spira et al. 1987).

Neurons isolated from adult insects retain many differentiated features under culture conditions, including membrane properties and action potential waveform (Newland et al. 1993), neurite appearance (Fig. 4.30) and some specificity in the selection of synaptic partners. The primary branching pattern characteristic of identified crustacean or insect neurons is more likely to be reexpressed in vitro if a length of axon is left attached when the neuron is isolated (Newland et al. 1993). Identified insect embryonic neurons appear to spontaneously differentiate characteristic monopolar and bipolar morphologies (Kim and Wu 1991). Similarly, identified leech neurons may reproduce characteristic major branches in culture (Acklin and Nicholls 1990). In contrast, snail neurons grow from rounded somata and fail to differentiate distinguishable axons and dendrites in isolated culture.

The question of what determines the distinction between axons and dendrites has been addressed in some vertebrate preparations. In some systems, differentiation results from contact with different substrate molecules (Thomson 1990). In other neurons, differentiation is an expression of intrinsic cellular properties (Dotti et al. 1988). In cultured embryonic hippocampal neurons, the MAP2 and tau markers chemically confirm the apparent anatomical differences in the neurites (Mattson et al. 1988b). During development, growth-associated protein, GAP-43, is found in axonal but not dendritic growth cones. In cultured neurons, the differentiation of neurites into axon and dendrites is correlated with the segregation of the GAP-43 into the axon (Goslin et al. 1988; Gordon-Weeks and Mansfield 1990; Gordon-Weeks 1991). GAP-43 is localized at the surface of the plasma membrane in association with the membrane skeletal fraction that contains fodrin, talin, and α-actinin and is therefore likely to be involved in the regulation of cell shape (Meiri and Gordon-Weeks 1990).

4.9 Designing Culture Conditions

Defining minimal conditions under which neurons can grow is just the beginning of the possibilities for studying neurons in vitro. Instead of presenting neurons with uniform conditions, the environment in the culture dish can be manipulated to test hypotheses about neural regeneration. The most common challenge is to culture neurons with other neurons or nonneural target cells.

Synapse formation and the influences of the target are discussed in the next chapter, so here we will here consider other influences on growth patterns.

By manipulating the molecules used on culture dishes, the role of substrates and soluble factors in adhesion and sprouting may be distinguished. Substrates such as polylysine may provide good adhesion but induce little sprouting, whereas others may also stimulate growth. In that case, the adhesive action of the substrate depends upon moities distinct from the growth-promoting part of the molecule (Goldberg et al. 1991). If a substrate that only provides good adhesion can be identified, soluble growth-promoting substances can be screened.

Laminins distributed along the path of neurite outgrowth promote substrate attachment and axonal outgrowth, among their several functions (Bixby 1989; Beck et al. 1990; Hunter et al. 1991; Reichardt and Tomaselli 1991; Venstrom and Reichardt 1993). The addition of laminin to neurons growing on polylysine triggers new growth in arrested growth cones of cultured rat sympathetic neurons (Goldberg et al. 1991). Laminin can also promote growth cone extension when added as a soluble factor (Chamak and Prochiantz 1989). Leech laminin supports both adhesion and sprouting of leech neurons, and has allowed culture in a defined medium (Masuda-Nakagawa et al. 1988).

The branches of regenerating cockroach giant interneurons exhibit different growth patterns in the ganglia and the connectives (Spira et al. 1987). This suggests that important elements in the microenvironment might be extracted from different CNS regions and tested in dissociated neuron culture.

The environment of isolated neurons has been manipulated by coating half of the substrate dish with conditioning medium (Wong et al. 1981) or presenting discontinuous choices between substrates (Grumbacher-Reinert 1989). The turning of an *Aplysia* growth cone to retain contact with a favored substrate involves significant contact by filopodia with the unfavored substrate followed by splitting of the growth cone at the border and survival of only the branch that extends on the preferred substratum (Goldberg and Schacher 1987; Goldberg et al. 1991).

Topographical discontinuities produced by photolithography can direct the growth of neurites (Kleinfeld et al. 1988; Wilson et al. 1991). The substrate can be constrained within narrow borders by covering a substrate-coated surface with a grid which shields a patterned distribution of the substrate from UV photolithography (Fromherz et al. 1991; Fromherz and Schaden 1994). The neurites maintain cytoplasmic contact with the patterned substrate, and branch points elicit bifurcation (Fig. 4.31). Encounters between identified neurons or with different substrates can thus be orchestrated.

The gradients of cues, such as those which guide pioneer axon growth along the proximal-distal axis of insect limbs (Norbeck et al. 1992), can be reproduced in culture. A gradient of membrane-associated cues can be created on the surface of a filter using the technique of Baier and Bonhoeffer (1991, 1992). This involves suction of a droplet of medium containing membrane fragments through a filter, with the droplet constrained under a tilted glass coverslip, so that as the fluid is drawn through, the droplet size declines and the remaining fluid is drawn toward the slide contact point, which will correspond to the

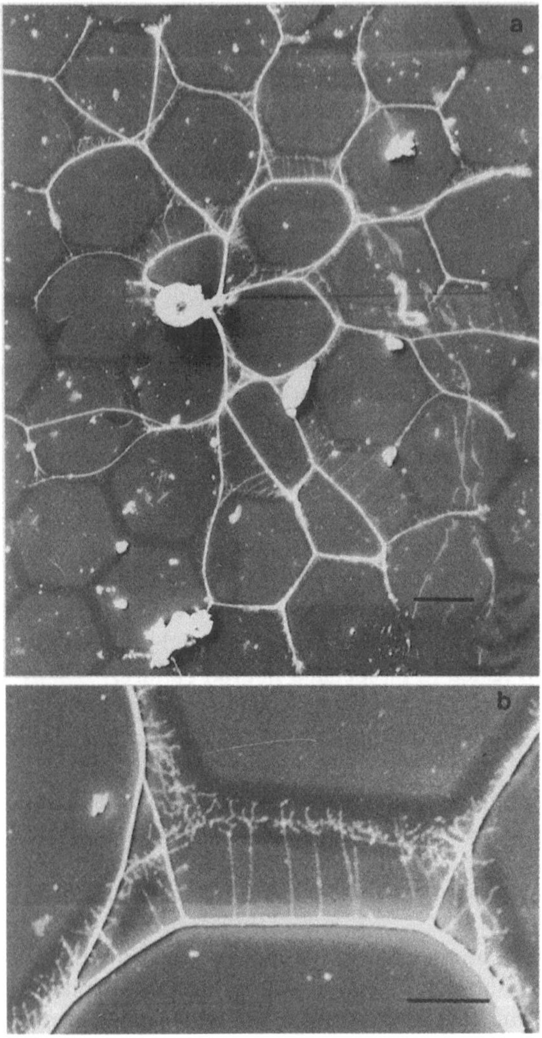

Fig. 4.31. Scanning electron micrograph of leech neurons. Guided outgrowth on hexagonal net of extracellular matrix/bovine serum albumen (ECM/BSA). **a** Overview; *bar* = 50 μm. **b** Detail; *bar* = 20 μm. Bent neurites are pulled off from bent lanes. Foot-like sites of adhesion are revealed at intervals of about 10 μm. The neurites are connected to these sites by filaments. (Fromherz et al. 1991)

region of highest concentration of the factor on the filter (Fig. 4.32). Soluble factors can also be presented in a gradient in polyethylene glycol.

Optical recording allows noninvasive monitoring of electrical activity in cultured neurons (Ross et al. 1987; Parsons et al. 1991) and could be optimized by growing neurons on grids, as described above.

The natural environment exists in three dimensions, whereas the typical culturing condition approximates a two-dimensional field. However, the availability of confocal microscopes should reduce our dependence upon two-dimensional neuron culture. A three-dimensional matrix of neurons suspended in a gel will allow a more normal morphology in response to localized cues, which can be more effectively controlled by the experimenter (Richards et al.

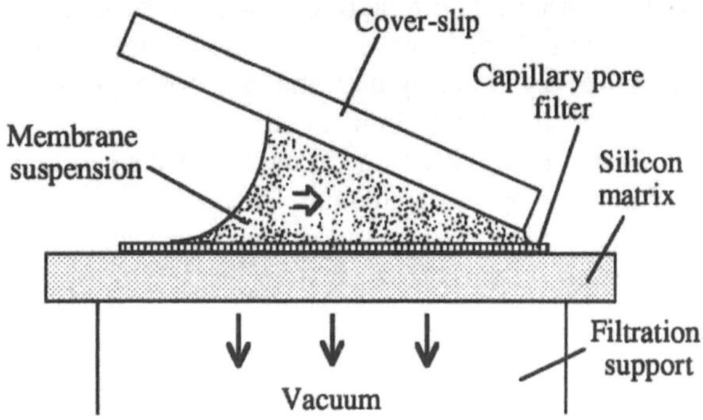

Fig. 4.32. Gradient production. An inclined glass coverslip holds a small drop of membrane suspension (100–500 μg of protein per milliliter) on top of a capillary pore filter. The filter is placed on a silicon matrix that is attached to a filtration support. When suction is applied (*three arrows*), the drop quickly decreases in volume, and because of its surface tension, contracts in the direction indicated by the *open arrow*. This continues until the underlying filter is plugged by membrane particles. These particles form a spatial gradient, which for technical reasons is of sigmoid shape. The superfluous suspension and the coverslip are removed. The second membrane species is added and sucked onto the filter. By plugging the remaining pores, these membranes form a symmetrical countergradient. (Baier and Bonhoffer 1992)

1983; Hamamoto et al. 1988). Embryonic dorsal root ganglion neurons have been grown in solid astrocytic tissue formed by packing the cells into a tube (Fawcett 1992). By combining some of these techniques with the three-dimensional approaches discussed above, it will be possible to present a neuron with something approximating the patterned environment it would naturally encounter, including choices between attracting and repelling substrates (Hitchcock and Raymond 1992).

4.10 Conclusions

In both developing and regenerating neurons, the growth cone's role is to lead the axon to an appropriate target. The senses it depends upon are primarily chemical, with receptor contact linking some molecules to the cytoskeleton and other receptors acting through ion channels and second messengers. The growth cone recoils from repulsive regions when contacts cause it to it withdraw filopodia or collapse; it is pulled into other regions when filopodial contacts promote extension.

An injured neuron immediately expresses a different pattern of genes. Some of these genes are undoubtedly involved in the reception of cues in the environment that will guide the growth cone; likewise, some substrate cues which are typically only seen during development are reexpressed in response to the neural injury. Whether or not the regenerating neuron undergoes a sequence of receptor-substrate interactions triggered by encounters with the environment is not known. Such formative alterations may occur in neurons during development, but regenerating neurons may not experience similar alterations in affinities until they reach their target.

Target recognition and synaptogenesis can transform the growth cone into a semistable synaptic element. Throughout life, use can modify functional features of the synapse and injury or target loss can set the cell free to seek new targets.

Chapter 5

Synapse Formation and Alteration During Regeneration

5.1 Introduction

Many interactions contribute to the formation of connections in developing and regenerating nervous systems. During critical periods in development, extrinsic influences interact with the timetable of intrinsic events to produce differentiated neurons and networks (Wall and Taghert 1991; Ushkaryov et al. 1992; White et al. 1992; Mimura 1993). Although the particular mix of intrinsic and extrinsic factors that guide differentiation and synapse formation differs for different animals and for different neurons in a given animal (Thomson 1990; Keshishian 1993), similarities in the molecular basis of synapse formation are becoming more and more apparent.

Two patterns of synaptic partner selection are illustrated by different neurons in the buccal ganglia of *Helisoma*. In the first pattern, neuron B 19 selects synaptic partners unerringly from among a range of potential synaptic targets, so that no pruning of the excessive connections occurs (Fig. 5.1). This pattern has been called "prospective specificity" by Haydon and Zoran (1994). In contrast, neuron B 5 initially forms many more synapses than are retained after the early period of exuberant connectivity (Haydon and Kater 1988), a "retrospective selectivity" pattern. One or another of these two patterns will be apparent throughout this chapter, and factors that regulate selectivity will be discussed.

5.1.1 Comparing Synaptogenesis in Regeneration and Development

If we compare circuit formation in developing and regenerating animals, it is apparent that axons regenerating in an adult nervous system are faced with different and sometimes more formidable tasks than those accomplished during development. Regeneration in an adult animal differs from development in a number of ways, including:

1. The distance to the target is greater (e.g., ten times as great in adult *Drosophila*; Keshishian et al. 1994). Guidance cues such as substrate gradients or target attractants must operate over greater distances, and the metabolic expenditure required for an adult-sized neurite to reach the target may sometimes exceed the growth potential of an adult neuron.

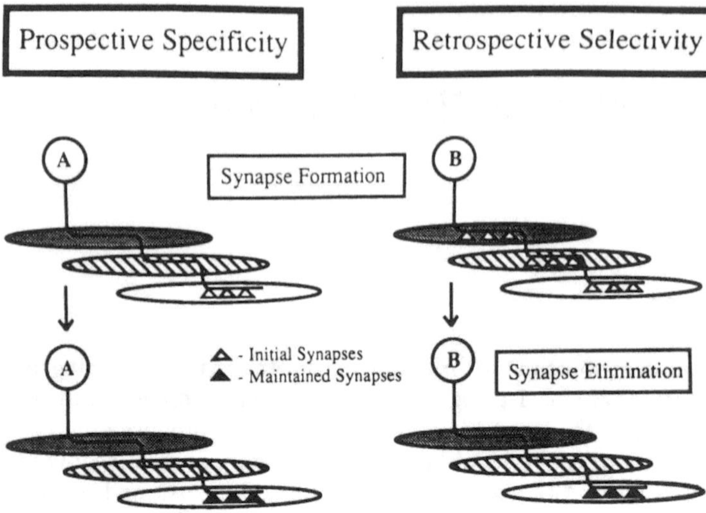

Fig. 5.1. Synaptic specificity involves the combination of *prospective* synapse formation and *retrospective* synapse elimination mechanisms. Specific patterns of neuronal connectivity are generated through the processes of initial synapse formation, based on a neuron's ability to recognize correct targets, and through the subsequent elimination of mistakes in connectivity. In this example, neuron *A* is highly specific in synapse formation, recognizing only a single muscle target as a synaptic prospect. This appropriate connection is maintained. Neuron *B*, on the other hand, is extremely nonselective in synapse formation, initially establishing many inappropriate connections. This neuron could either recognize all targets as potential partners or could simply possess no target-discriminating capabilities. In retrospective selectivity, appropriate connections are maintained through activity-dependent or activity-independent mechanisms of novel synapse elimination. (Haydon and Zoran 1994)

2. Development often involves cells that appear transiently during development to guide axonal growth, and there is no evidence for the reappearance of these cells in adult neurite regeneration.
3. The orderly timing of neurite arrival at a target site is lost when an injury sets many axons into synchronous growth. However, the outcome may not be as different from that seen in development as might have been expected (Chiba and Murphey 1991).
4. Differences in connectivity between normal development and regeneration may be attributable to the loss of critical-period refinement of connectivity or the decline in plasticity of adult neurons.
5. Neurons may differentiate as a result of their encounters with the target during development, whereas neurons rarely revert to an undifferentiated state after target loss. In this sense, developing and regenerating neurons are not the same cells and cannot be expected to behave in the same ways.

Despite these differences, regenerating nervous systems of many animals are able to duplicate or approximate original circuitry in vivo. Most neural regeneration involves responses made by the surviving, differentiated elements navigating through differentiated tissue. For such cells, the reexpression of cues required for neuronal differentiation would be superfluous, because regenerating neurons typically retain a sense of identity. Nevertheless, regenerating neurons operating in the complex environment of an adult animal depend upon some form of guidance to locate appropriate targets.

5.1.2 Novel Connectivity in Regenerated Nervous Systems

The connections initially formed by regenerating systems, both in vivo and in vitro, are not necessarily restricted to the repair of connections established during development. Following a given lesion, each identified neuron will recreate the same pattern of connections with some neurons and not others. If the neuron follows the retrospective pattern (Fig. 5.1), its synapses may include "correct" partners found in control animals, appropriate partners (contralateral homologues or neurons that have synergistic actions), or inappropriate connections with no known behavioral function. For example, leech neurons form connections that are not necessarily the same as those found in the animal (Fuchs et al. 1981; Aréchiga et al. 1986; Vyklicky and Nicholls 1988).

The sign of the synaptic connection may differ from that seen in controls. The sign of the synapses formed in culture between *Aplysia* abdominal ganglion neuron L10 and its target neurons is different from the control pattern (Kleinfeld et al. 1990). Such a difference can theoretically be due to different transmitter release or different receptor-mediated conductances. For instance, there are instances of steroid-induced changes in transmitter expression during insect metamorphosis (Tublitz 1993). Which transmitter such neurons would express if removed from the adult and grown in culture has not been tested.

Even when the sign of the synapse is the same, the functional properties of the synapse may not be the same. The crustacean neuromuscular system has provided a wealth of information about homosynaptic changes in transmission as a function of activity. When fatigue and facilitation at neuromuscular synapses was compared in control and regenerated crayfish claws, resistance to fatigue at the synapses was abnormally low in the regenerated claws (Stewart and Atwood 1992).

Novel connections are gradually eliminated after normal synaptic contacts form but may persist if appropriate target neurons have been eliminated (Bulloch et al. 1984; Denburg 1985; Pallas and Hoy 1986; Baptista and Macagno 1988). Changes in the strength of existing connections may also occur in response to injury, often increasing the efficacy of a stimulus that was formerly inadequate to tigger a response. In these ways, new, workable networks can be created.

5.1.3 Reduced Plasticity in Regenerating Neurons

Gao and Macagno (1988) discuss the critical period for refining connectivity based on feedback from innervation and competition with other neurons during leech development. Leech neurons, and the neurons of most animals, have a greater capacity to interact with targets and undergo refinement during development than during regeneration. For instance, S cell behavior is different during development, apparently obeying different stop growth cues (Muller et al. 1987). Similarly, embryonic snail neurons display greater plasticity than adult neurons in their responses to serotonin (Goldberg et al. 1990).

Survivable injuries that involve the loss of neurons pose special challenges for the neurons seeking the missing cells as synaptic partners and for the reestablishment of function in the nervous system as a whole. After neuron loss, remaining neurons need to be able to accept appropriate, alternative targets if they are to make the best repairs possible under the circumstances.

To the extent that the period of rapid regrowth of processes following injury fails to allow correct or compensatory circuitry to form, the process of regeneration is defective. A loss of flexibility may constitute a major difference between neural development and regeneration in some animals.

5.1.4 Circuits Successfully Reformed in Vivo and in Vitro

Despite the various ways in which regeneration can fail, many examples of near-perfect repair are known. The recovery of function in circuits after disruption of the network argues that both the wiring specificity and the nature of the synapses have been restored. An example of such repair is seen in the regeneration of the respiratory network of the snail *Helisoma*.

The three-neuron network which controls respiratory behavior in *Lymnaea* and *Helisoma* has been studied after connections are broken by axotomy (Roger et al. 1992). The neurons are located in three different ganglia, LPeD1 in the left pedal ganglion, RPeD1 in the right pedal ganglion, and VD4 in the visceral ganglion. This separation facilitates interruption of connections in an acutely isolated nervous system that can subsequently be maintained for four days in vitro. Intracellular iontophoresis of Lucifer yellow dye reveals multiple neurites originating from the stumps, some of which project in directions appropriate for reestablishment of connections. Electrophysiological recordings reveal connections of the same sign as in the control circuit (Fig. 5.2).

A respiratory circuit neuron transplanted into a host visceral ganglion from which the corresponding neuron (VD4) has been removed can establish connections and restore function in the respiratory circuit (Syed et al. 1992a).

When neurons of the *Lymnaea* respiratory circuit are isolated and cultured together, the respiratory circuit is recreated in vitro (Syed et al. 1990). Chemical synapses characteristic of the intact circuit form and generate alternating bursts similar to the in vivo pattern (Syed et al. 1990; Fig. 5.3). The rhythmic pattern is

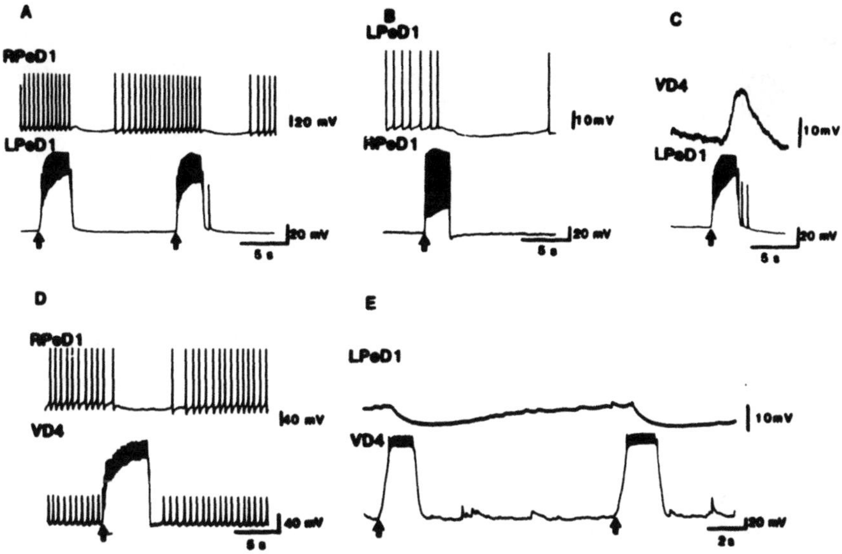

Fig. 5.2. Reestablishment of synaptic connections in an interneuronal network of *Helisoma*. Intracellular recordings demonstrate the formation of specific synaptic contacts between appropriate neurons following axonal regeneration. Both LPeD1 and RPeD1 reestablished their reciprocal inhibitory synapses (A, B). The excitatory connection of LPeD1 onto *VD4* was also reestablished after 4 days (C). Similarly, the inhibitory connections between VD4 and the other two identified interneurons (*RPeD1, LPeD1*) were also restored after 4 days in ganglion culture (D). (Roger et al. 1992)

not produced by any single pair of the neurons, but requires the contribution of all three (Syed et al. 1993).

This feat of circuit reconstruction in vitro demonstrates that differentiated neurons, in the absence of cues from the environment, possess the intrinsic information necessary to recreate circuitry, including the strength and particular features of synaptic interactions crucial to circuit function. If a circuit is hypothesized to have certain characteristics, based on in vivo recordings, and displays the same features in vitro, this is the best possible test of the hypothesis. Furthermore, the isolated circuit can be subjected to further manipulation, such as the effects of modulators.

5.2 The Road to Recovery

5.2.1 Behavior of Injured Animals

Behavior is not always a good index of neural regeneration. Early observations of leech responses to neural injury revealed that behavioral recovery is not dependent upon the regeneration of lesioned connections (Jansen and Nicholls

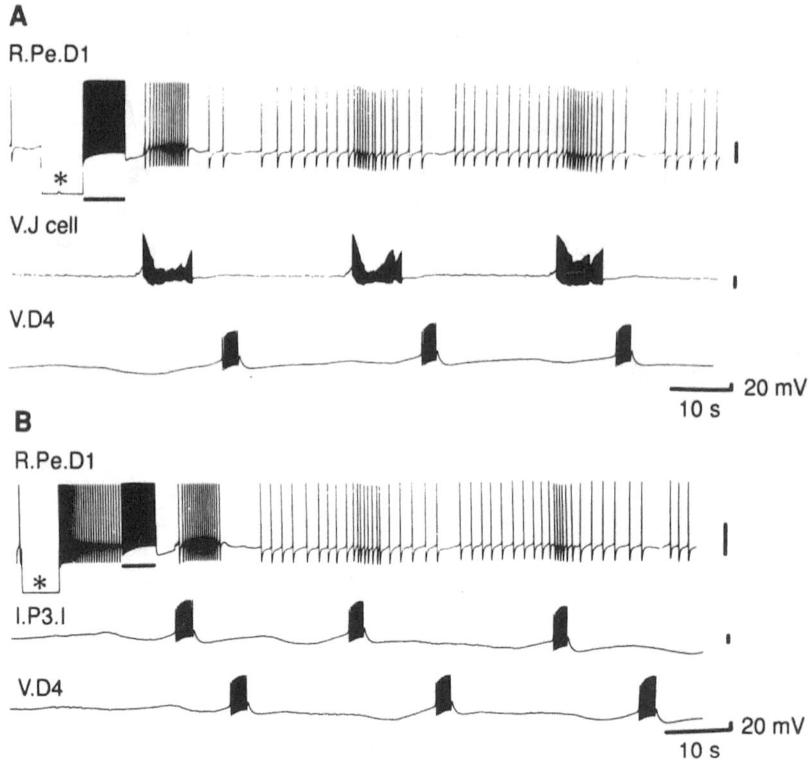

Fig. 5.3. Intracellular recordings from respiratory interneurons and the giant dopamine cell of *Lymnaea* **A** in situ and **B** in vitro. **A** Stimulated respiratory rhythm recorded from the isolated brain preparation. Indirect evidence for the occurrence of activity in IP3I was obtained from its follower, visceral J-cell (VJ cell). Direct recordings were made from RPeD1 and VD4. Injection of hyperpolarizing current into RPeD1 (*asterisk*) revealed a lack of passage of hyperpolarizing current between these interneurons. Detailed analysis of this three-cell circuit both in vivo and in vitro also showed a lack of electrical coupling between these neurons (not shown). In contrast, when depolarizing current was injected into RPeD1 (*at bar*), it initiated IP3I activity (as recorded from V.J-cell) while inhibiting VD4. The activation of IP3I in turn excited the giant dopamine cell and the previously hyperpolarized VJ-cell while inhibiting VD4. Upon recovery from inhibition by IP3I, V.D4 fired a burst of action potentials, and the cycle was spontaneously repeated. **B** Both the respiratory interneurons and the giant dopamine cell were isolated in culture and allowed to extend neurites. They were then impaled with glass microelectrodes. Injection of hyperpolarizing current into the RPeD1 (*asterisk*) revealed a lack of electrical coupling. Injection of strong depolarizing current (*at bar*) inhibited both respiratory interneurons. The direct interaction between RPeD1 and VD4 can be seen more clearly during the train of action potentials that precedes the strong depolarization of RPeD1 (*at bar*). IP3I recovered from this inhibition and fired a burst of action potentials due to the postinhibitory rebound. This activation of IP3I in turn excited RPeD1 and caused the further inhibition of VD4. Similarly, VD4 recovered from the inhibition by IP3I and fired a burst of action potentials inhibiting both RPeD1 and IP3I. Thus, an alternating pattern of bursting activity was initiated by RPeD1 that continued for several cycles before reaching a quiescent state (n = 8). This in vitro pattern of activity mimics the respiratory cycle seen in both isolated and semi-intact preparations. (Syed et al. 1990)

1972; Jansen et al. 1974). Changes in existing connections and the formation of new connections are seen after connective transection, and leeches thereby regain the ability to generate coordinated movements in the absence of regrowth across the lesion. Similar adjustments in circuits occur in response to deafferentation (see Sect. 5.6).

Even when it reflects the state of neural repair, behavior is a crude assay. We know from the small diameter of regenerating axons and the behavior of recently formed synapses that the connections are weak and labile at first and mature over a period of days to weeks (Macagno et al. 1985; Glanzman et al. 1989). They may never regain their original strength (for reasons described in Sect. 5.2.4).

After neural injury, an animal's behavior may pass through recognizable stages that reflect the condition of the underlying circuitry. Immediately after extensive CNS damage, animals may appear stimulated or depressed, depending on whether the injury activates or blocks remaining circuitry. In operations performed under anesthesia, animals are spared this initial response phase.

The snail *Melampus* has been subjected to a range of extensive neural ablations under anesthesia. Soon after recovering, snails withdraw into the shell, exhibiting little spontaneous behavior. When spontaneous behavior reappears and responses to stimulation can be observed, the deficits related to the lesion can be evaluated. Prior to behavioral recovery, however, is a stage marked by hyperexcitability (Moffett and Snyder 1985; Snyder and Moffett 1987, 1990). For instance, in the first week following cerebral commissure transection, snails respond with abnormally rapid and prolonged withdrawal to minor stimulation such as shadows or air puffs. This generally disappears within 2 weeks, by which time bilateral retraction and correlated anatomical evidence of repair often indicate that new projections connect the cerebral ganglia. Within 1–2 months, the snails achieve a stable state characterized by partial or complete restoration of behavior and a return to normal levels of sensitivity.

In *Aplysia*, relatively minor neural injury caused by a pinch lesion of the skin results in site-specific sensitization of defensive reflexes (Walters 1987a,b). The excitability of sensory neurons is elevated by neuromodulatory factors, which sum with the intrinsic effects of high-frequency firing in the injured or strongly stimulated cells (Fig. 3.9). Facilitation of the tail withdrawal pathway after neural injury is attributable both to a lower sensory neuron threshold and to enhanced monosynaptic connections onto the motor neurons. The resulting ease of activation of this behavior would serve to protect injured body parts during healing (Walters 1991).

5.2.2 Connectivity Changes and Hypersensitivity

Injury unleashes a latent propensity for sprouting and the potential to form novel and altered connectivity as well as to reestablish normal connections. The hyperexcitability observed in *Melampus* may have a parallel in the injury-

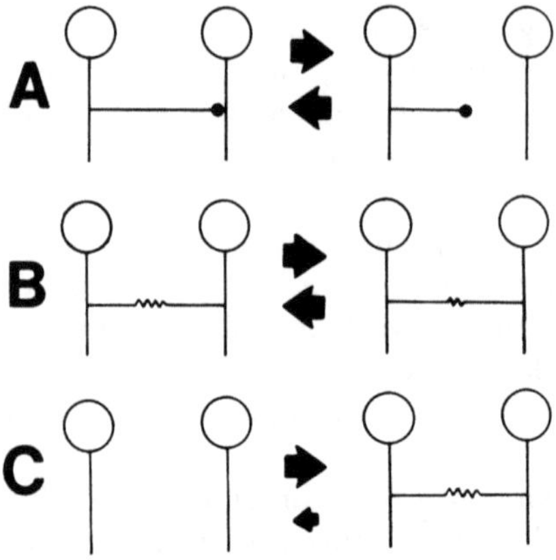

Fig. 5.4. Schematic summary of the types of synaptic plasticity observed to date in the buccal ganglia of *Helisoma*. **A** Chemical inhibitory synapses between premotor neurons (cyberchrons) and the protractor motor neurons exhibit a transient loss of efficacy, which is restored by 2–4 weeks of culture. **B** Electrical coupling between the same groups of neurons is increased with predictable functional consequences. Indirect evidence suggests that this is also transient. **C** Selective formation of new, stable electrical connections. *Black ball* contact represents chemical inhibitory synapse; *shorter resister element* represents lower resistance, i.e., stronger electrical synapse. *Arrows* in **A** and **B** indicate that processes depicted are apparently reversible. In contrast, the *unequal arrows* in **C** indicate that some of the new electrical connections are transient, whereas others, e.g., 5–5 in paired buccal ganglia and 5–4 in single ganglia, are stable. (Bulloch and Kater 1982)

induced changes in *Helisoma* connectivity (Bulloch and Kater 1982; Hadley et al. 1983; Murphy et al. 1984; Bulloch 1985). In the isolated *Helisoma* buccal ganglia, reduction in the efficacy of inhibitory chemical synapses and an increase in electrical coupling are seen (Fig. 5.4; Bulloch and Kater 1982). A predictable result is an increase in excitability. Formation of electrical synapses between neurons in *Helisoma* is favored if both potential partners are actively extending neurites (Hadley et al. 1983).

The widespread electrical connections have anatomical correlates in the extensive sprouting documented following injury. In *Lymnaea*, crushing the axon of giant interneuron RPeD1 elicits extensive novel neuritic projections (Fig. 5.5) but results in reestablishment of normal connections, even when the neurons travel by novel routes to reach their targets (Allison and Benjamin 1985; Benjamin and Allison 1985).

5.2.3 Remodeling of Connections

The widespread sprouting characteristic of gastropod responses to injury and the resulting formation of novel connections is subject to gradual remodeling. The novel electrical connections that appear in snail nervous systems in the early stages of regeneration in some animals may disappear in a matter of days to weeks, following a time course similar to recovery from hypersensitivity (Fig. 5.6). Spontaneous disappearance of the connection between *Helisoma* buccal neurons 4 and 5 occurs within 48 hours (Bulloch and Kater 1981, 1982). This electrical connection is maladaptive, linking neurons that normally activate antagonistic muscles in different phases of the buccal rhythm. Other novel electrical synapses between neurons 5 and retractor motor neurons peak within 1

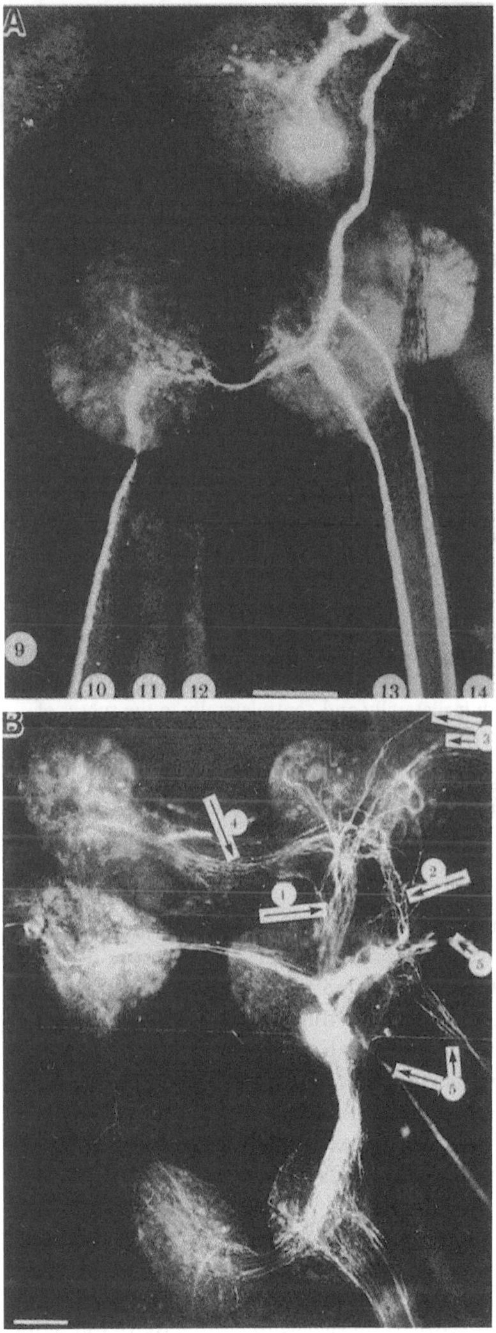

Fig. 5.5. A Lucifer yellow fill of normal RPeD1 in isolated brains, demonstrating the anatomy of the interneuron. Major invariant axon branches occur in the anal nerve (*10*) and in the internal (*13*) and external (*14*) right parietal nerves. Several fine neurites also extend a short distance into the genital nerve (*12*). **B** Anatomy of regenerated RPeD1 revealed by Lucifer yellow injection. Regenerated neuron 28 days after a lesion of the right pleural-parietal connective. Neurites have extended in large numbers posteriorly across the crush site into the right parietal and visceral ganglia and also anteriorly into regions of the CNS not normally invaded by RPeD1 processes. Scale *bar* = 200 μm. (Allison and Benjamin 1985)

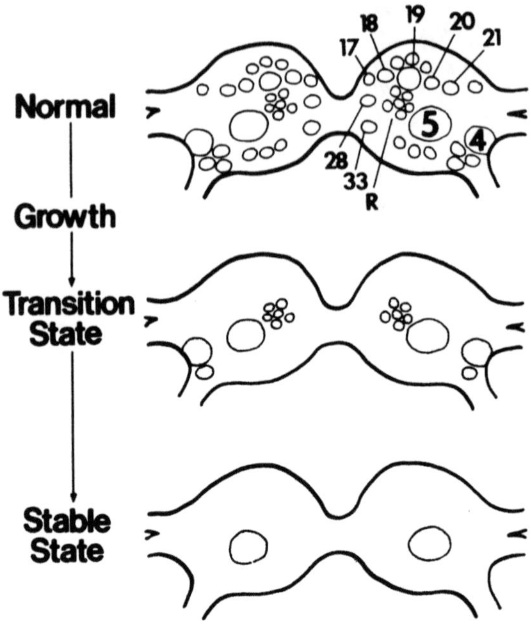

Fig. 5.6. Summary of the new electrical connections observed in the buccal ganglia of *Helisoma*. Neurons assayed for coupling to neuron *5* are depicted in the *upper diagram* and include neuron *4* (salivary neuroeffector), the protractor motoneurons (*17–21*) and the retractor motoneurons (*28, 33*) and a group here shown as *R*. Additionally, a number of previously unidentified neurons (*unlabeled circles*), which were recognizable as groups rather than individuals, were tested. In acutely isolated ganglia, neuron 5 was not coupled either to its contralateral partner nor to any other neuron tested. (Thirty cell pairs were assayed in ten preparations.) When neuritic growth was evoked, however, a number of new electrical connections were routinely found after short culture durations, neuron 5 apparently being connected both ipsilaterally and contralaterally to the subset of neurons illustrated in the *middle diagram*. Most of these new connections were eliminated on further culture (1–2 weeks), only 5–5 coupling being stable (*lower diagram*). This survey for new connections was performed on ganglia mainly after either 1, 2, 7–8, or 13–14 days of culture and involved more than 260 dual intracellular recordings in 80 preparations. (Bulloch and Kater 1982)

day and slowly decline in strength during the following week, disappearing by 2 weeks. The most stable of the transient connections is between the homologous neurons 5; this gradually diminishes over 8 weeks, although anatomical projections thought to underlie the connections may still be present (Cohan et al. 1987a). Unlike the novel connections, normal connections are stable when tested at 10 weeks, and are probably permanent.

Possible consequences of increased excitability in the early stages of recovery include facilitation of withdrawal reflexes that compensate for weakened or deficient motor or sensory capabilities. The changes observed in *Aplysia* would fall under this category. Such hyperexcitability is by no means a universal response. In leeches, injury alters sensory-to-motor circuits by substituting predominantly inhibitory for predominantly excitatory effects, potentially depressing behavior (Jansen and Nicholls 1972). Perhaps the best generalization about stimulation of growth and widespread increases in connectivity following injury is that these provide a mechanism for bridging lesions.

Altered connectivity produced following injury in leeches and insects may be relatively permanent (Jansen and Nicholls 1972; Jansen et al. 1974; Pallas and Hoy 1986; Lakes et al. 1990). Persistent altered connectivity may interfere with establishment of correct connections or may permanently distort the animal's behavioral repertoire. Similarly, in the lamprey, repair following spinal injury is reflected in anatomical and behavioral recovery, but also some nonfunctional growth patterns (Cohen and Baker 1988; McClellan 1992).

5.2.4 Consequences of Deafferentation

Loss of sensory inputs is one major category of neural injury. Sensory systems are often subject to experience-related adjustment of synaptic strength, and their plasticity is reflected in their responses to injury (Bailey and Chen 1989; Chiba and Murphey 1991; Bailey et al. 1992). In the period before a regenerating sensory projection reaches its destination, vacated input sites can be occupied by other sensory cells, and central circuitry controlling contralateral projections or unrelated circuits can occupy vacated sites. Deafferentation typically alters the strength and/or distribution of connections among remaining elements (Lnenicka and Murphey 1989). This is evidence that competitive interactions are important in determining and maintaining the strength of synaptic inputs.

5.2.4.1 Effects of Sensory Deprivation

The Insect Cercal System.
One of the best studied systems in orthopteran insects (cricket, grasshopper, cockroach) is the system of connections between the mechanosensory cells of the paired posterior structures called cerci and the giant interneurons upon which the cercal afferents synapse. The sensory hairs of the paired cerci convey information on the presence and directionality of wind and low-frequency sounds. The synapses of the sensory neurons project in a topographical pattern on the giant interneurons, which control orientation responses and coordinated escape reflexes (Murphey et al. 1980).

Modulation of synaptic efficacy and even changes in the afferent's target occur during growth. Sensory neurons are added to the cerci at each molt (Edwards and Palka 1974). As the number of inputs increases, the strength of the synapses of individual cercal afferents declines and some of the sensory cells shift inputs from their original target onto another interneuron. The consequence is that input-output properties of the giant interneuron are maintained at a constant level during development (Chiba et al. 1988).

A lesson apparent from the insect cercal system is that compensatory responses to deafferentation may impose a persistent bias on a neural circuit. Even though the sensory neuron synapses onto the giant interneurons regenerate after the sensory axons are cut (Bacon and Murphey 1984; Shepherd et al. 1988; Chiba and Murphey 1991), anatomical and physiological bias may remain. For instance, if cerci are removed in a larval stage, they regrow and the afferents

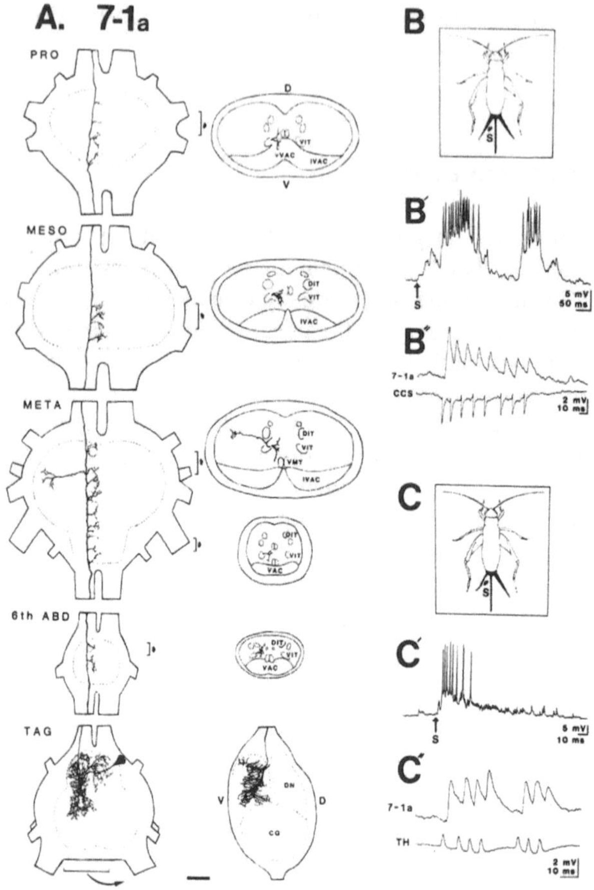

Fig. 5.7. Analysis of synaptic connections in the CNS of the cricket following transplantation of peripheral structures containing sensory neurons. Interneuron *7-1a* received synaptic input from abdominal sensory neurons. **A** Anatomy of interneuron *7-1a* in the abdominal and thoracic ganglia. Note that since the anatomy of this neuron is very similar in each of the four free abdominal ganglia, only one (sixth abdominal) ganglion is shown. Transverse (thoracic and 6th abdominal) and sagittal terminal abdominal ganglion (*TAG*) views of each ganglion are shown to the *right*. *Brackets* indicate the extent of each ganglion used to reconstruct the sectional views. *D* Dorsal; *DIT* dorsal intermediate tract; *DMT* dorsal median tract; *MVT* median ventral tract; *V* ventral; *VAC* ventral association center; *aVAC* anterior VAC; *lVAC* lateral VAC; *mVAC* ventralmost VAC; *VIT* ventral intermediate tract; *VMT* ventral median tract. *Scale bar* = 100 μm. **B** Receptive field of neuron *7-1a* in a control animal. The response of *7-1a* to tactile stimulation of the cercus ipsilateral to its large dendrite (*arrow*) is shown in **B'**. *Arrow* indicates approximate onset of stimulation. All recordings are from *7-1a*'s large dendritic segment in the *TAG*. **B"** Simultaneous recordings from *7-1a* and a cercal campaniform sensilla (cS) sensory neuron. **C** Receptive field of *7-1a* includes the ectopic leg (*arrow*) in an experimental animal. **C'** Response of *7-1a* to tactile stimulation of the ectopic leg ipsilateral to its large dendrite. **C"** Simultaneous recordings from neuron *7-1a* and a tactile hair sensory neuron on the ectopic leg. (Killian et al. 1993)

form appropriate connections, even if regeneration is delayed (Edwards and Palka 1971). Even temporary deafferentation increases the strengths of contralateral and noncercal vibratory inputs and inhibitory inputs from the contralateral cercus (Palka and Edwards 1974). Removal of a cricket's cercus during the juvenile stage results in the expansion of the dendritic arbors of the remaining receptor cells, including a normally bilaterally projecting neuron (Shepherd and Murphey 1986). Even where functional parameters are restored by regeneration, the structure of the giant interneurons may reflect periods of deafferentation (Murphey et al. 1976). Selective reduction of the input region is also observed in higher-order interneurons after first-order interneurons are ablated (Casagrand and Ritzman 1991).

Experiments with transplanted sensory hairs and appendages show that sensory cells regenerate along modality-specific paths and can recognize correct targets or substitute appropriate ones. In the cricket *Acheta domesticus*, cercal afferents normally innervate interneuron 7–1a (Fig. 5.7; Killian et al. 1993). This interneuron does not normally receive inputs from leg tactile afferents, but if a leg is transplanted into a cercal position, the tactile afferents of the leg grow in the same tract normally followed by cercal afferents and synapse onto 7–1a. Similar appropriate substitutions are made when cerci are transplanted to the cricket's thorax. The afferents grow along the tactile sensory cell pathway rather than the proprioceptive tract and provide tactile information relevant for the region of the body where they are positioned. This situation is remindful of the growth pattern of leech ectopic afferents from transplanted sex structures, which successfully follow positional cues into the CNS in non-sex ganglia (Passani et al. 1991).

Mutants of *Periplaneta americana* have been discovered that differ in the number of filiform hairs on their cerci (Bacon and Blagburn 1992). These hairs are innervated by single, identifiable sensory neurons, such as L and M, and in "space invader", competition provided by this additional "SI" neuron for target space significantly reduces the size of neuron L's EPSP (Fig. 5.8, cf. A and B). If both neurons are ablated, the inputs of neuron M expand (Fig. 5.8C). These observations reinforce the conclusion that postsynaptic cells restrict the total input that they receive, and that sensory cells compete with one another for space or input strength.

The Crayfish Tailfan Inputs.

The mechanosensory receptors on the crayfish tailfan represent a sensory system similar to the orthopteran cercal system. These afferents synapse onto giant interneuron A (Kennedy 1971; Strandburg and Krasne 1985). The synapses are normally all ipsilateral, but the interneuron can be forced to accept synapses from the contralateral side if the ipsilateral afferents are cut or functionally silenced by a covering of glue (Krasne 1987). Observations in this system support the conclusion that the target plays a role in regulating total synaptic input and activity is important for maintaining and regulating afferent input strength.

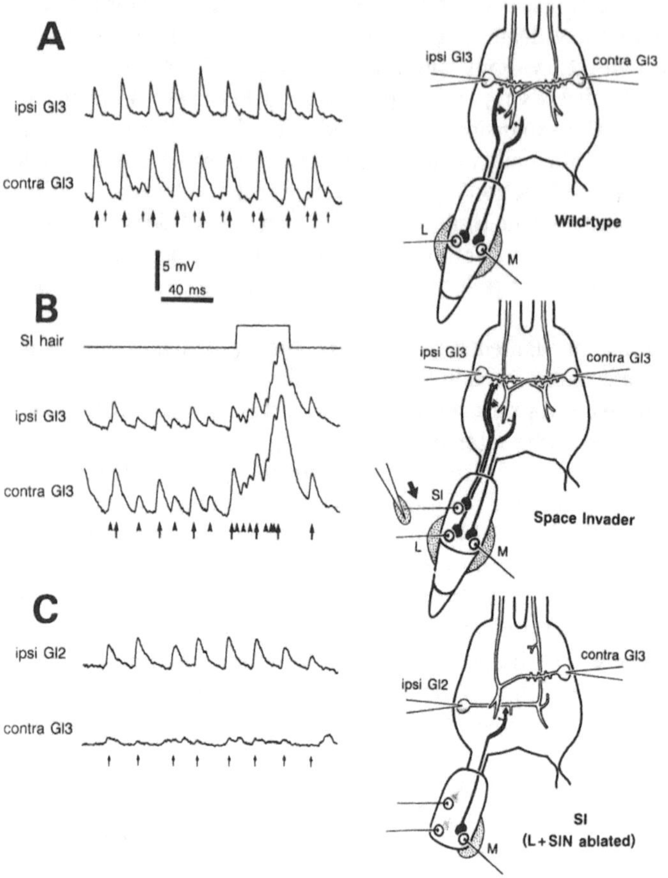

Fig. 5.8. Intracellular recording in the terminal ganglion shows that supernumerary afferents make synaptic connections. The right-hand cercal nerve was crushed in all of these experiments and the L- and M- filiform hairs were immobilized with petroleum jelly (*stippled*) which causes regular, tonic spiking in the afferents. The diagram to the *right* of each trace depicts the experimental situation. **A** Simultaneous recording in the ipsi- and contralateal G13s in a wild-type animal. Large EPSPs (*large arrows*) occur synchronously in both G13s; these results are from the input of the regularly spiking L afferent. Small EPSPs in the contralateral G13 only (*small arrows*) arise from the M afferent. **B** In *SI* individuals, simultaneous recording in ipsi- and contralateral G13s reveals a set of additional synchronous EPSPs (*arrowheads*) arising from *SIN*. Their identity is confirmed by moving the SI hair (displacement of *top trace*), inducing a burst of summating SIN EPSPs in both GIs, during which the larger L EPSPs (*arrows*) are visible. It is not possible to detect the small M EPSPs in contralateral G13 amongst the larger EPSPs. **C** To allow the measurement of the small M EPSPs in the preparation shown in **B**, the L and SIN somata were ablated and the ipsilateral G12 was impaled. The large EPSPs in G12 can be used as markers for the occurrence of M EPSPs in contralateral G13 (*small arrows*). (Bacon and Blagburn 1992)

Insect Auditory System.
The auditory system of insects is another sensory system that must occasionally adjust to injury (Kalmring 1975; Huber and Thorson 1985). The hearing organ of locusts and crickets is located in tibia of the first thoracic limbs; sensory afferents project ipsilaterally to the frontal acoustic neuropil of the metathoracic ganglia. Destruction of the sensory cells in one tympanic organ of *Locusta migratoria* results in sprouting of the intact contralateral afferents, which cross the midline within 5 days and reinnervate the auditory interneurons after 20 days (Lakes et al. 1990). The other response to deafferentation is sprouting of some deafferented interneurons into the contralateral neuropil. Apparently this adjustment alone is adequate to attract afferent input and restore function in *Schistocerca*, a related locust in which afferent projections do not cross the midline in larvae or adults subjected to unilateral deafferentation (Lakes et al. 1990).

In contrast to locusts, crickets regenerate the tympanic organ along with other structures if the leg is amputatated at the femur-tibia joint (Huber 1987). At least some of the alterations in interneuronal morphology are retained after regeneration of sensory afferents in crickets (Pallas and Hoy 1986; Schildberger et al. 1986). However, even monaural crickets regain their ability to respond to the direction of a sound (Schmitz et al. 1988; Schmitz 1989). Lakes (1990) compared several orthopteran species and concluded that competence to regenerate peripheral structures such as auditory receptor cells was not correlated with central plasticity.

5.2.4.2 Central Compensation for Sensory Deficits

Everyone is familiar with the ability of insects and spiders to locomote despite the loss of legs or the tattering of wings. This capacity clearly demonstrates that these animals are far more complex than the wind-up toys their behavior sometimes suggests. For example, the plasticity of the locust flight system allows these insects to generate normal behavior despite significant deletion of control system components (Büschges and Pearson 1991). In locusts, sensory structures called hindwing tegulae provide excitatory inputs to interneurons that drive wing elevation. Removing the inputs from the hindwing tegulae decreases the wingbeat frequency, increases the interval between elevation and depression of the wings, and alters phase relationships (Büschges and Pearson 1991). Normal flight behavior gradually returns within 2 weeks, whether or not the insects fly during that interval.

In contrast to the hindwing tegulae, the forewing tegulae normally have minor and quite variable input to the flight interneurons and their ablation has little effect on the flight pattern. However, if they are ablated in the recovered insects, the flight pattern is disrupted just as severely as it was following the initial hindwing tegulae ablation. Electrical recordings show that the interneurons of insects that have recovered from ablation of hindwing tegulae are receiving more and stronger inputs from the forewing sensory cells (Fig. 5.9; Büschges et al. 1992a,b). The interneurons have altered dendritic trees which

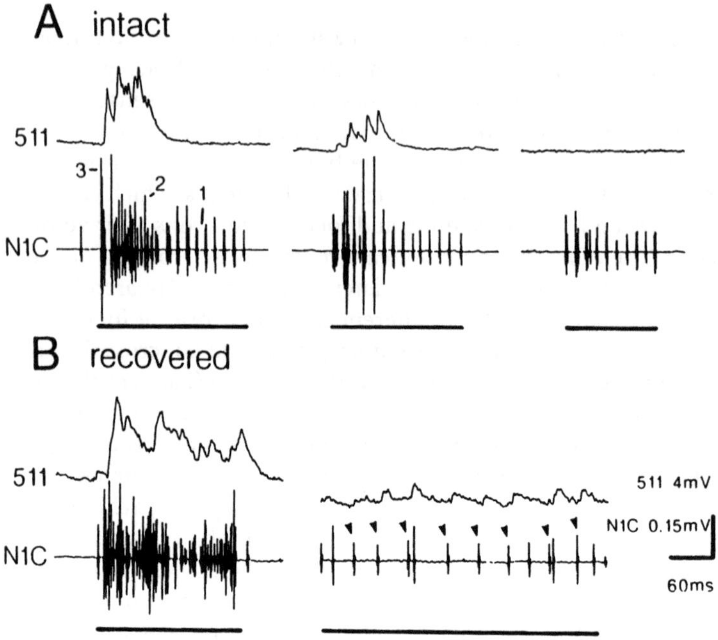

Fig. 5.9. Intracellular recording from interneuron 511 during mechanical stimulation of one forewing tegula in **A** an intact and **B** a recovered animal. Afferent activity was monitored by extracellularly recording from nerve 1C (*N1C*). Note that in **A**, the small forewing tegula afferents did not evoke unitary EPSPs in interneuron *511*, whereas in **B**, almost all small afferents induced EPSPs (*arrowheads*). The *black bars* indicate the approximate time of mechanical stimulation of the forewing tegula. (Büschges et al. 1992)

correlate with these changes (Fig. 5.10). These alterations allow the locusts to substitute alternative sensory input in order to maintain an excitatory level crucial for the proper operation of a motor control circuit. Such adjustments illustrate the utility of a system based upon competitive interactions.

Many other examples of compensation for deafferentation exist. For instance, long-term stability of the crab (*Carcinus maenas*) ventilatory central pattern generator is maintained after unilateral deafferentiation (Wilkins 1994). Anatomical adjustments to the loss of afferent inputs are seen, but the adjustments in connectivity or membrane properties that underlie the recovery of autogenic rhythmicity in this system remain to be elucidated. Perhaps the alteration in burst propensities in cultured spiny lobster neurons represents a similar phenomenon (Turrigiano et al. 1994).

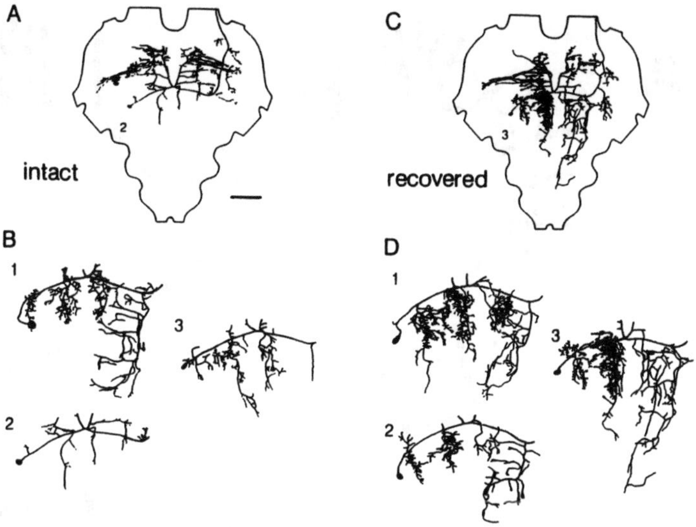

Fig. 5.10. Comparison of the morphology of interneuron 566 in the metathoracic ganglion of intact and recovered animals. A Morphology of interneuron 566 in an intact animal. C Morphology of interneuron 566 in a recovered animal. In B and D, partial drawings of interneurons 566 are shown for three intact and three recovered animals, respectively. These drawings show an increase in arborizations of interneurons 566 posterior to the primary neurite. (Büschges et al. 1992)

5.2.5 Compensation for Ablated Ganglia

The CNS of the snail *Melampus* has been subjected to severe ablations, including removal of many of the individual ganglia (Price 1977; Moffett and Snyder 1985; Moffett unpubl.). The remaining neurons are able to grow through a site of ganglion ablation and establish sufficient connections to support behavioral recovery (Fig. 5.11). Behaviors such as tentacle withdrawal in response to both ipsilateral and contralateral tentacle stimulation require synaptic connections and efferent control from the remaining cerebral ganglion (Moffett and Snyder 1985). Following excision of one cerebral ganglion, the metacerebral giant neuron in the remaining ganglion extends its projections across the buccal commissure and out through nerves and the connective of the contralateral side (Moffett and Ridgway 1988a).

Locomotion is coordinated by the paired pedal ganglia (Moffett 1990). The locomotor pattern generators in the paired pedal ganglia are normally linked through the pedal commissure, and cutting this commissure causes the animal to locomote utilizing the two sides of the foot in asynchronous wave patterns, but this abnormal pattern disappears with regeneration of the pedal commissure (Snyder and Moffett 1990). Compensatory adjustments in locomotion are also seen in *Melampus* following pedal ganglion removal. After one pedal gan-

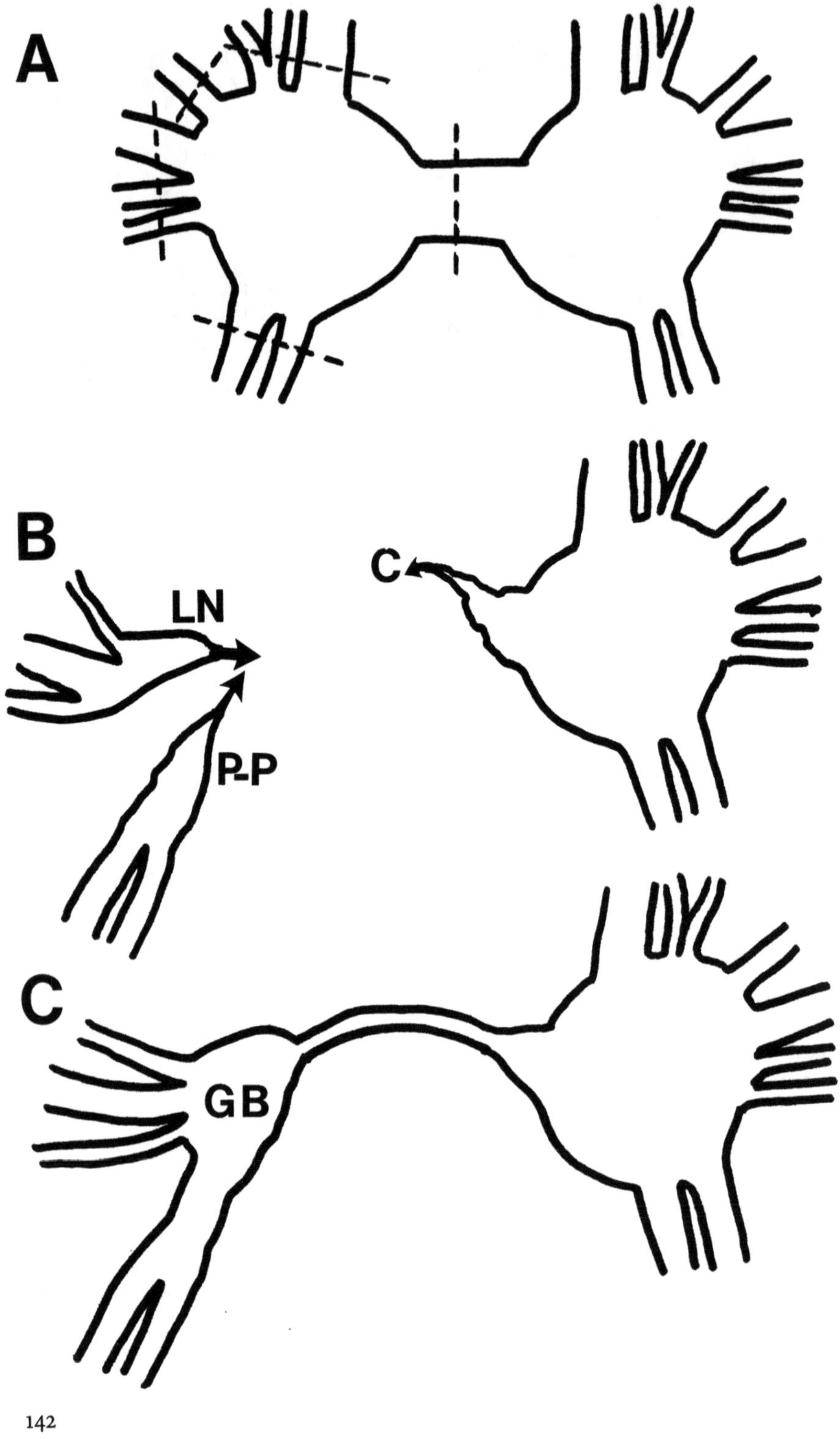

A
B
LN
P-P
C
C
GB

glion is removed, projections traced in backfills of the regenerating pedal nerves reveal more neurons than normal projecting from the contralateral pedal and the other central ganglia (Snyder and Moffett 1990). Both sides of the foot are driven by the pedal locomotor rhythm of the remaining ganglion. Central reorganization includes a pathway from the cerebral ganglion on the side of the ablated pedal ganglion to the contralateral pedal ganglion, as can be demonstrated by the persistence of locomotor behavior in recovered animals after the connectives on the intact side are cut (Snyder and Moffett 1990).

5.2.6 Ganglion Regeneration

Within 6–9 months after ganglion ablation, regrowth of ganglion buds that include neurons has been documented in a third of the *Melampus* that were allowed to regenerate beyond 6 months (Moffett and Snyder 1985; Moffett and Ridgway 1988b). How the ganglion buds, consisting of sheath elements, neuropil, and tracts, acquire a rind of neurons has not been determined. The developmental origin of neurons in the body wall has been established for *Aplysia* (McAllister et al. 1983; Jacob 1984; Cash and Carew 1989; Hickmott and Carew 1991) and the addition of neurons to the ganglia throughout the juvenile period strongly suggests that the body wall is a likely source for replacement neurons. Neurogenesis is known to continue into adulthood in *Melampus* (May et al. 1987), so it is also possible that the new neurons are just some of the newly generated ones which move in from the body wall or spread from the other ganglia. However, the ultrastructural study of regenerated *Melampus* ganglia (Moffett and Austin 1982) suggests the presence of large, unique neurons. The ganglion regeneration study needs to be repeated now that uniquely identified neurons have been mapped by neurotransmitter antibodies. Short-term regeneration reveals neurons in the bud with reactivity to a variety of neurotransmitter antibodies. Additionally, some neurons are seen in commissures and connectives, which is not the case in control nervous systems (Moffett, unpubl.).

◀ **Fig. 5.11.** Cerebral ganglion regeneration in *Melampus*. **A** A pair of intact cerebral ganglia; *dashed lines* indicate sites of transections required for removal of the left ganglion from the snail. **B** The pattern of regrowth that forms a bridge through the site of the missing ganglion. The size of the *arrows* indicates the relative contribution of growth from the pedal and pleural connectives (*P-P*), the labial nerve complex (*LN*) and the contralateral cerebral ganglion (*C*). **C** The formation of a ganglion bud (*GB*) within approximately 1 month supports behavioral recovery, but central connections and neuropil continue to develop, and in some animals, ganglion regeneration through the addition of neurons is seen

5.3 How Neurons Select Targets

The prospective and retrospective processes of target selection both involve selectivity, but differ in how much leeway is allowed in the initial steps in the process (Fig. 5.1). Even when prospective selection normally operates, experimental manipulations can sometimes reveal a hierarchy of alternative synaptic partners.

5.3.1 Limitation of Choices by Pathway Cues

One way that neuronal circuits can be specified is for neurons to be programmed to follow cues set down along the path to their targets. This is the "blue print hypothesis" originally proposed for guidance of axons developing and regenerating in the newt spinal cord (Singer et al. 1979).

In development, the pathway, particularly for pioneer neurons, can be encouraged by permissive cues and guided by attractive or repulsive cues. Neurons following the pioneers have the simpler task of growth governed by selective fasciculation with axons travelling the same path. The growth of most regenerating neurites probably resembles the demands on the follower neurons more closely than the trail-blazing actions of pioneer neurons. The nerve sheath, distal processes of the same neuron, or the target are likely to release attractive cues. Repulsive secreted and substrate cues thus far identified in regeneration act in the mammalian CNS to block rather than guide axons.

In many nervous systems, growth cones that are able to locate and grow along the remnants of the degenerating nerve tend to arrive at the target and reinnervate it reliabily. Leech sensory and motoneurons with somata in the CNS grow to the periphery after axotomy and reestablish normal peripheral fields (Van Essen and Jansen 1977). There appears to be little competition, and hyperinnervation occurs in the case of duplicated neurons (Kuffler and Muller 1974). Growth may be more limited by intrinsic restraints on the extent of dendritic branching than by responses to features of the terrain, but where neuron elimination opens up additional territory, leech neurons can expand their field of innervation (Macagno et al. 1989).

When the mechanism of target selection involves following a pathway, neurons are typically quite successful in regenerating connections with appropriate central or peripheral targets (see Fig. 2.10). However, experimental manipulation has revealed that some synapses form on the cells at the end of the path without evidence of selectivity. For instance, leech motoneurons (anterior pagoda cells) that normally innervate ventral skin will readily accept dorsal skin if a rotated patch of skin is encountered (Zhang and Nicholls 1983). In contrast, leech T-cells regenerating within the ganglia normally follow the processes of other T cells closely, but regeneration along incorrect pathways can also lead to correct target selection (Macagno et al. 1985). In that case, the target must be sought, either by following cues or by randon extension of processes; once contacted, the target must be recognized.

5.3.2 Cell-Adhesion Compatibility in Synaptogenesis

The cell-adhesion systems described in Chapter 4 are not only important for getting neurons into proximity with their targets. Adhesion between potential partners may allow the cells to achieve the intimate contact necessary for the next stage of interactions.

Some regions within the nervous system may be identified by a regional distribution of substrate or surface markers. In *Aplysia*, an element of selectivity relating to the ganglion of origin is apparent when isolated neurons are plated together. Electrical synapses form among buccal neurons or bag cell neurons, but electrical synapses are not readily detected between bag and buccal neurons, despite an extensive overlap of processes and the ability of buccal neurons to form an inhibitory chemical synapse with bag cells (Bodmer et al. 1984). Extending the approach to include additional tests of abdominal and pleural ganglion neurons reveals only weak heteroganglionic electrical synapses

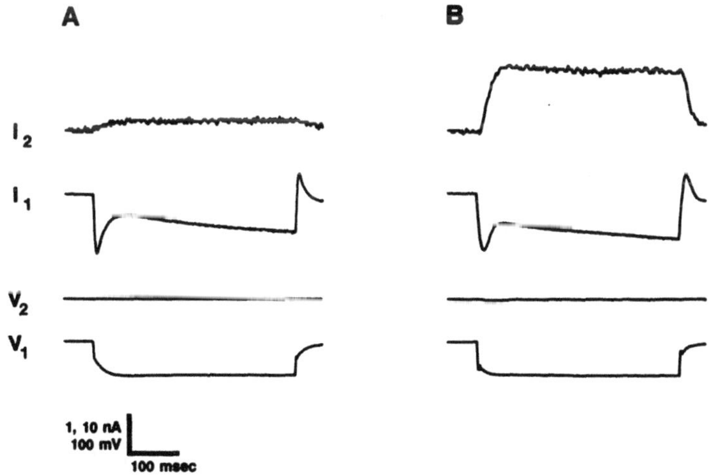

Fig. 5.12. Modulation of junctional conductance by Con A in cultured *Aplysia* neurons. Voltage-clamp records for a buccal-pleural pair (heteroganglionic) held at -70 mV. **A** Voltage pulse applied to one cell (V_1) activated nonjunctional membrane currents in that cell (I_1); there was no change in the holding potential of the postsynaptic cell (V_2), and the resulting junctional current (I_2) indicates a G_j of 2 nS. **B** After perfusion of the cells for 10.5 h with 10^{-7} M Con A, the junctional current (I_2) had increased an order of magnitude, reflecting a G_j of 18 nS. The nonjunctional current (I_1) was essentially unchanged, indicating no significant change in membrane input impedance. The voltage clamps were adjusted for stability at the expense of high-frequency response, and thus the voltage responses were relatively slow. (Carrow and Levitan 1989)

(Carrow and Levitan 1989). Ganglion selectivity is masked and weak interactions strengthened when the plant lectin Con A is added to the medium, which suggests that a glycoprotein plays a role in conferring specificity (Fig. 5.12; Lin and Levitan 1987; Carrow and Levitan 1989; Chap. 4).

Other neurons can also control adhesion molecule expression and the distribution of neurons. Interactions involving adhesion are apparent when *Aplysia* sensory neurons are cultured in the presence or absence of their target cell, neuron L7. The sensory cells, when cultured alone, grow over and along one another freely, but when cultured with a motoneuron, avoid contact and make segregated synapses on the motoneurons (Glanzman et al. 1991). This pattern makes sense in terms of what neurons need to do in the transition between pathfinding and synapse formation. The tendency of sensory neurons to fasciculate with one another is appropriate for forming a nerve, but when growing processes come into the vicinity of a potential target, the fasciculation with other sensory neurons is disadvantageous.

The tendency for sensory neurons to form synapses onto L_7 in vivo and in vitro is promoted by the addition of serotonin (Zhu et al. 1994). Its effect on the sensory cells is to down-regulate the expression of apCAM (*Aplysia* cell-adhesion molecules) in those cells, making the accumulations of apCAM on presumptive synaptic sites on the target cell relatively more attractive (Fig. 5.13). The reduction of fasciculation between the sensory neurons that results from exposure to serotonin is remindful of the changes seen in developing leech neurons as they enter a ganglion and defasciculate to form synapses, only to subsequently fasciculate into several discrete tracts under the control of different adhesion molecules (see Figs. 4.22–4.24).

The processes of some cultured *Aplysia* neurons, whether normal synaptic partners or not, may exhibit a tendency to fasciculate in culture. This tendency is a good predictor of the likelihood that they will form chemical synapses, whereas neurons that do not fasciculate typically fail to form synapses (Hawver and Schacher 1993). Some pairs that are compatible in terms of transmitter-receptor responses (e.g. L10 and right upper quadrant cells) are reluctant to fasciculate, and form synapses only 20% of the time. When the tendency of these neurons that avoid overlap is counteracted by intertwining the processes, forced contact synapses form 77% of the time, compared with 88% for preferred partner pairings (Hawver and Schacher 1993). Thus selectivity is related to the degree of membrane overlap, which is promoted by compatible surface interactions. Right-upper-quadrant neurons are known to lack apCAM (Keller and Schacher 1990; Mayford et al. 1992), which may explain the low frequency of synapses between these cells and L10.

In conclusion, the surface adhesion molecules, which are widely but not universally distributed among neurons in a given nervous system, may represent a screen which typically prevents some neurons from coming close enough to form synapses. In insects, the adhesion molecule fasciclin III is involved in neuronal fasciculation and also in the formation of the synapses (Halpern et al. 1991). There is no high degree of selectivity in the operation of these adhesion systems, but their distribution on the postsynaptic cell may also play a role in

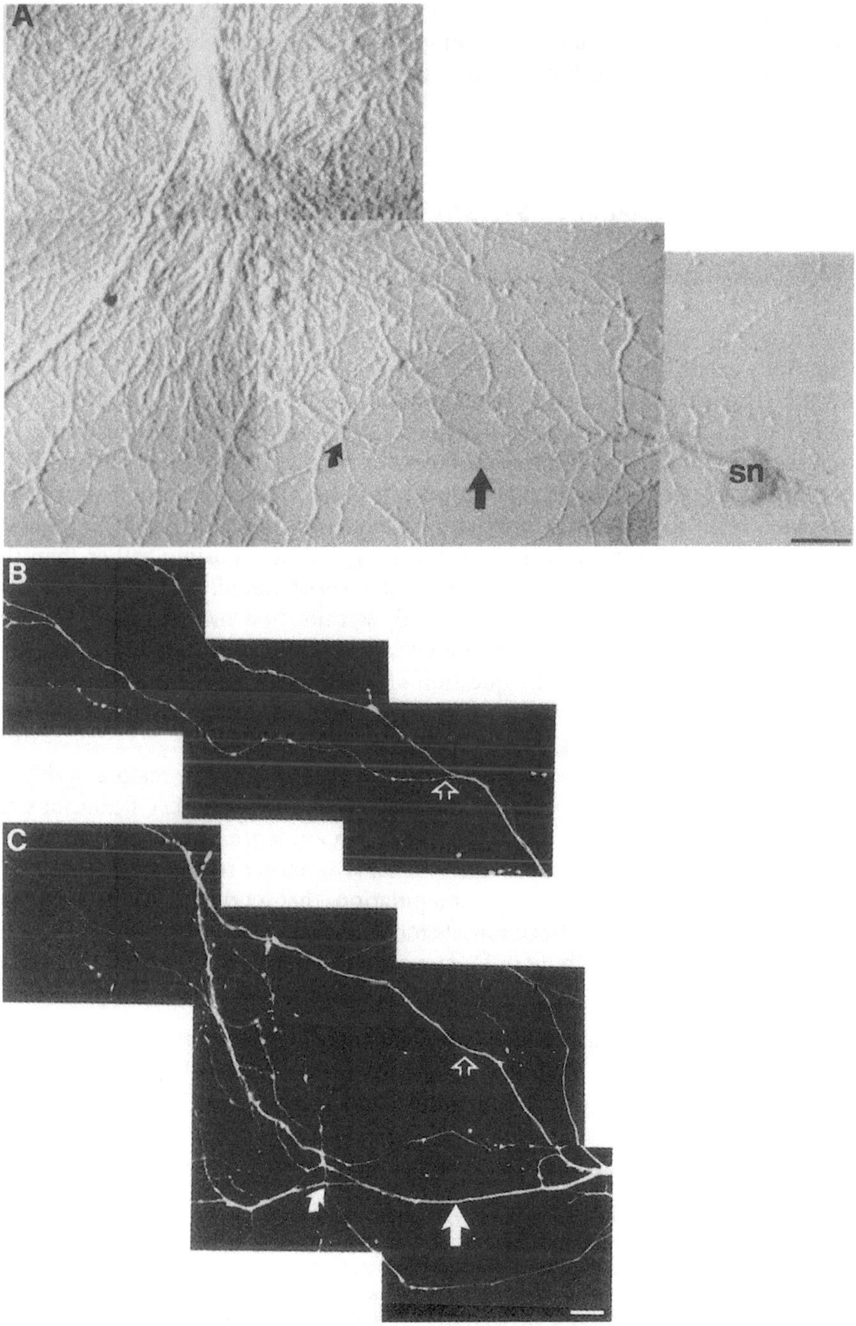
A
sn
B
C

the mechanics of synapse site selection. Some neurons will not form synapses with inappropriate partners even though they have compatible surface adhesion molecules (Haydon and Kater 1988; Haydon and Zoran 1989).

5.3.3 Specific Recognition Systems

5.3.3.1 Specificity in Arthropod Neuromuscular Innervation

Selectivity by Crayfish Motoneurons.
Naturally existing differences in the innervation of serially homologous muscles of the crayfish abdomen have been employed in experiments that explore the origin of the specificity. Six neurons innervate the superficial muscles in each of the second and third abdominal segments of crayfish, but the neurons exhibit a reversed pattern of muscle distribution in these two segments. If nerves are diverted from their normal innervation field to the adjacent segment, neurons project in the pattern associated with their ganglion of origin rather than the pattern typical of that abdominal segment (Worden et al. 1988). This suggests that when neurons regenerate successfully within their own segment, they are obeying neuron-intrinsic recognition cues for specifically labeled muscles. Kirk and Govind (1992) discuss the question of whether the initial differentiation of this pattern of neuromuscular specificity is dependent or independent of contact with the target during development.

Regeneration of six crayfish motoneurons is able to restore a highly position-dependent distribution of the endings on the superficial flexor muscles (Ely and Vélez 1982; Hunt and Vélez 1982). No evidence for the operation of a gradient in the recreation of the innervation is apparent (Goransson et al. 1988). If the target area is limited, a manipulation that corresponds to extensive muscle damage, the superficial muscle motoneurons may differ in their capacity to respond, but all contribute to functional reinnervation (Clement et al. 1983).

Cockroach Motoneuron Regeneration.
Control innervation patterns achieved during development in the cockroach neuromuscular system have been compared with the results of regeneration (Pearson and Bradley 1972; Whitington 1979). A coxal muscle that is normally

◀ **Fig. 5.13.** 5-HT changes the structure of regenerating sensory neurites interacting with distal neurites of L7. **A** Nomarski contrast photograph of SN-L7 culture on day 3. The sensory cell was plated about 500 μm from the stump of one of the proximal axons of the target motoneuron L7. **B** Epifluorescent photograph of all regenerated sensory neurites contacting the distal neurites of L7 on day 2. **C** Epifluorescent photograph of regenerated sensory neurites contacting distal neurites of L7 24 h later and following applications of 5-HT. New sensory neurites branch extensively and form numerous varicosities, exhibiting a pattern of growth that is more typical of sensory neurites contacting the proximal axons of L7. The *arrows* indicate reference points. Scale *bars*: **A** = 70 μm; **B** and **C** = 30 μm. (Zhu et al. 1994)

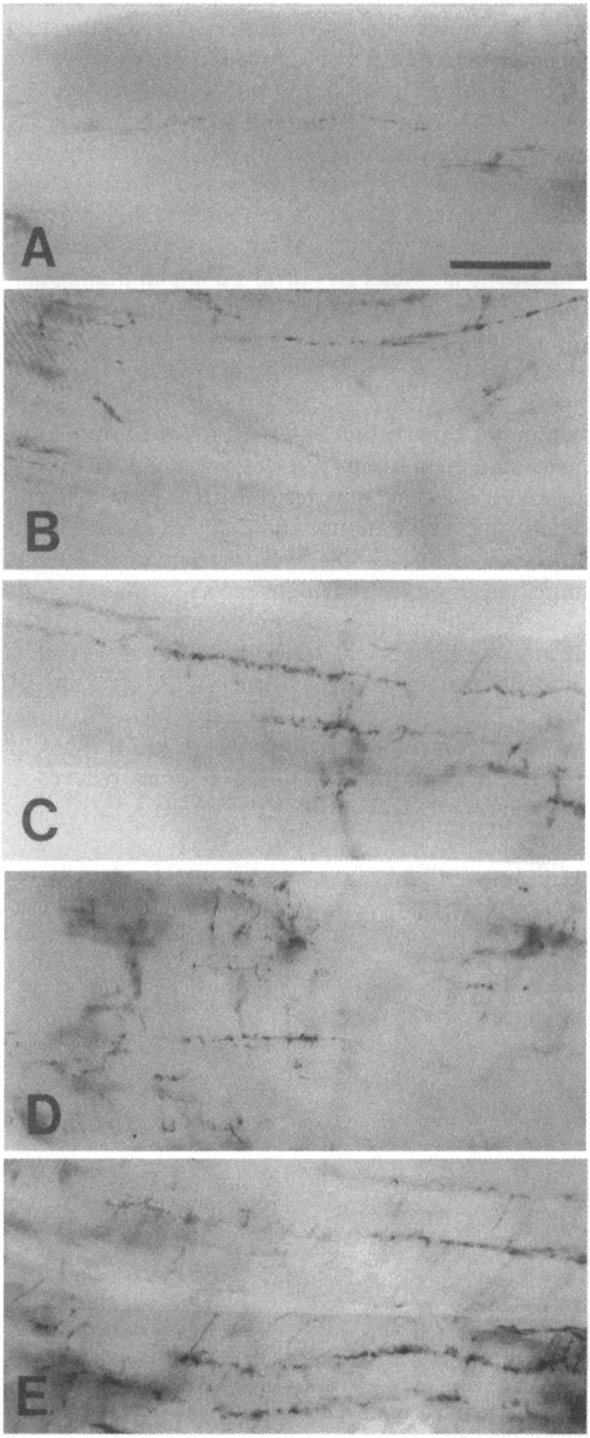

Fig. 5.14. Target recognition in the regeneration of cockroach motor axons. The monoclonal antibody NSS-2A-labeled axons project into appropriate target muscle 177e at **A** 20 days; **B** 40 days; **C** 60 days; **D** 80 days; **E** 100 days after nerve crush. Scale *bar* = 100 μm in **A** and applies to **B-E.** (Denburg and Caldwell 1992)

innervated by a single neuron can receive inputs from nearly 30 regenerating motoneurons in early stages of regeneration (Denburg 1988). Hyperinnervation is followed by the refinement of connectivity based on cell-cell recognition (Denburg et al. 1977; Denburg 1982). Plant lectins bind to subsets of the leg muscles in a pattern that corresponds to the innervation pattern (Denburg et al. 1983a,b) and motoneuron subsets are labeled by antibodies that selectively bind to glycoprotein antigens on axon terminals (Denburg et al. 1987). This information probably reflects the molecular specificity used by neurons to recognize particular muscles during development. Molecules involved in pathfinding and target recognition are typically expressed transiently, but antibodies reveal that many of these molecules are reexpressed in regeneration. Examples include the regeneration in newts (Fekete and Brockes 1987) and cockroaches (Denburg and Norbeck 1989).

Refinement of cockroach motor innervation has been followed over time. When regenerating axons grow into the periphery, they grow along the muscle fibers of both correct and incorrect target muscles, but only project lateral branches onto the correct fibers (Fig. 5.14; Denburg and Caldwell 1992). Formation of these collateral branches is the morphological correlate of selectivity, and coincides with the elimination of projections onto incorrect muscles (Fig. 5.15).

The many examples in insects of abnormal trajectories that nevertheless lead finally to correct target identification provide evidence of target selectivity (Kämper and Murphey 1987; Murphey et al. 1983a,b). Further evidence that recognition rather than competition governs target innervation is provided by neuron elimination experiments in which the remaining neurons, relieved of potential competition, do not retain incorrect connections on denervated mus-

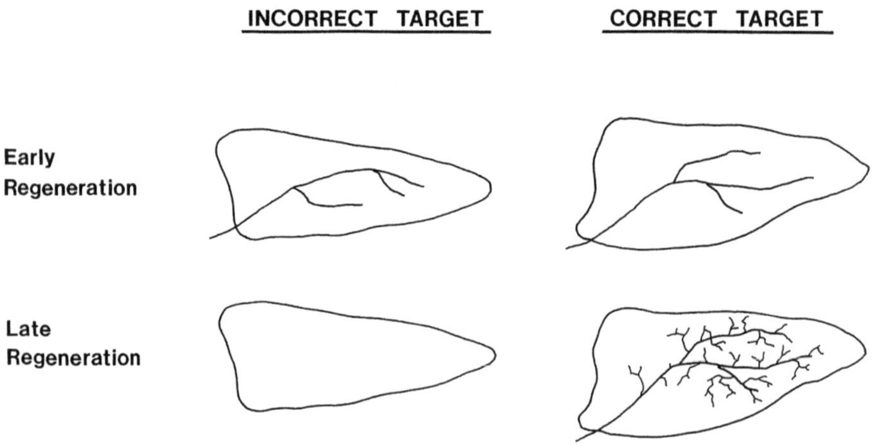

Fig.5.15. Summary diagram of the patterns of growth of the regenerating axon branches. At early stages of regeneration, the axon branches grow parallel to the muscle fibers in both correct and incorrect target muscles. At late stages of regeneration, collateral branches have grown circumferentially around muscle fibers in the correct target while axon branches in the incorrect target are eliminated. (Denburg and Caldwell 1992)

cles (Denburg et al. 1988). (Note that the response to neuron elimination is different from the response to muscle elimination: Denburg 1985, and see below.)

In discussing adjustments within a single neuron during the period of process refinement, Denburg and Caldwell (1992) suggest that the tendency to balance production of muscle-specific dendritic sprouts with retraction of incorrect projections is consistent with the sibling neurite bias hypothesis, according to which, growth is differentially distributed among neurites on the basis of feedback obtained from targets (Smalheiser and Crain 1984). Contact with a preferred target can differentially influence the transport of internal resources, thereby determining which branches are retained and which are eliminated (Goldberg and Schacher 1987). This redistribution of resources reflects competition between processes of an individual neuron, as contrasted with the competition between neurons.

Drosophila Motor Innervation.

Development of motor innervation in *Drosophila* indicates that the filopodia of motoneurons sample all the muscle fibers in their path and unerringly select the preferred fibers (Keshishian et al. 1994). Duplications or deletions

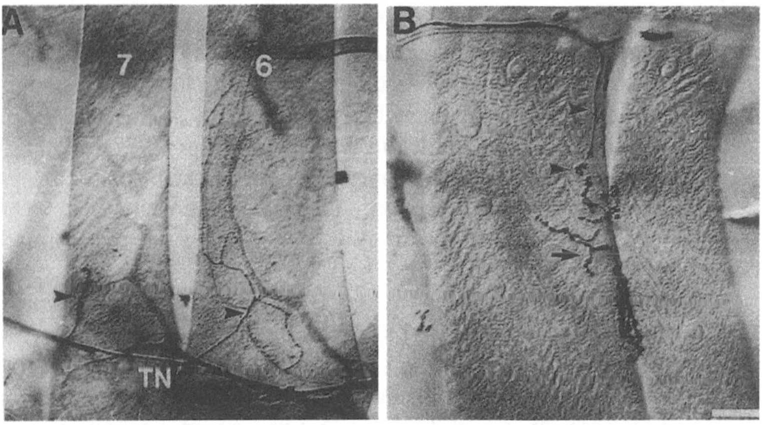

Fig. 5.16. Neuromuscular development in *Drosophila*. Both denervation and reduced neural activity lead to the appearance of ectopic foreign motor endings. **A** When the RP3 motoneuron is deleted by single-cell laser ablation in intact embryos, the denervated muscle fibers *7* and *6* receive foreign, ectopically placed inputs (*arrowheads*), shown here in a third instar larva. Note that the cleft between muscle fibers *7* and *6* is missing its native ending (cf. **B**). Ectopic motor endings were observed in 84% of denervated muscle fibers (n = 40 hemisegments) compared with 7.5% in innervated muscle fibers (n = 40 hemisegments). Similar ectopic motor endings arise after genetic or pharmacological treatments, which suppress presynaptic electrical activity by reducing sodium channel expression. **B** The appearance of an ectopic motor ending (*arrowheads*) in a nap^{tsl} larva, emerging from the transverse nerve (*TN*). The *arrow* indicates the native motor ending, located in the cleft between muscle fibers *7* and *6*. Animals were shifted from the permissive (18°C) to the restrictive temperature (37°C) for 6 h each day. The frequency of ectopic contacts in the activity blocked mutants increased to 30% (n = 84 hemisegments) compared with 6% (n = 69) in temperature-shifted, wild-type controls. Peroxidase-labeled anti-HRP immunocytochemistry. *Scale bar* = 25 µm.(Keshishian et al. 1994)

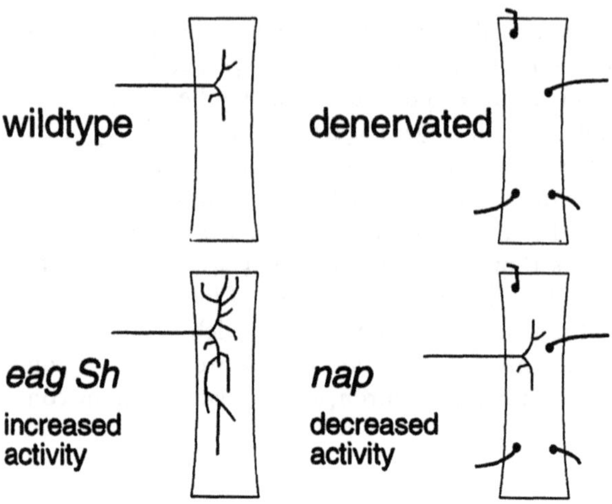

Fig. 5.17. Summary of the effects in *Drosophila* of denervation and activity mutations. In eag Sh larvae, with increased electrical activity, motoneuron branch arbors increase in size and complexity. By contrast, larvae with reduced neural activity (nap^{tsl}) or larvae where the motoneuron had been deleted receive increased numbers of foreign ectopically placed inputs. (Keshishian et al. 1994)

of muscle fibers show that the selection is muscle-specific, and a neuron may fail to innervate any muscle if its preferred target is eliminated (Chiba et al. 1993). In other instances, laser lesions of muscles result in ectopic innervation by motoneurons, and a hierarchy of acceptability may be apparent in the target muscles selected after the normal target is eliminated (Cash et al. 1992).

In contrast to the evidence for unerring innervation described above, some insect motoneurons apparently undergo refinement to produce the adult pattern (e.g. Myers et al. 1990). Sink and Whitington (1991a,b) found that if the correct muscles are not present, incorrect innervation established during the brief period of neurite expansion persists. Likewise, in the cockroach, if the normal target muscles are eliminated, withdrawal from incorrect target muscles never occurs (Denburg 1985).

Denervation of *Drosophila* muscles leads to sprouting and innervation by adjacent motoneurons (Fig. 5.16A; Keshishian et al. 1994). The absence of electrical activity in motoneurons, either as the result of mutations (Fig. 5.16B) or pharmacological block, leads to a similar invasion of the deprived muscle fiber. As discussed by Keshishian et al. (1994), although neuronal activity is not required for target selection, it is important for maintaining the pattern established during embryogenesis and for the balanced elaboration of synaptic contacts (Fig. 5.17).

5.3.3.2 Specific Connections Made by Transplanted Neurons

Transplantation of insect limbs to ectopic locations and the specific reinnervation of appropriate substitute targets shows that the neuron-for-muscle recognition system is distributed in a serially homologous pattern (Young 1972, 1973; Westin and Camhi 1975). The specificity of reinnervation is relatively independent of the pathway followed; rather it is the result of selective retention of correct choices after contact with appropriate and inappropriate muscles.

Cross-species transplantation of sensory neurons of crickets show that they can relate to their new position in the host body and project appropriate inputs on interneuronal targets (Kämper and Murphey 1987). This shows that the molecular basis of neuron identification is conserved across species, as are many of the other molecular cues used in development and regeneration. The ability of a neuron transplanted from a different species of snail to restore circuit function similarly attests to the robustness of the molecular recognition systems employed in synaptogenesis (Syed et al. 1992b).

Connections between leech Retzius (R) cells and P cells are faithfully regenerated in vivo and in vitro (Fuchs et al. 1982). A mature R cell implanted in a host ganglion will form electrical connections with the host R cell and will receive a novel excitatory chemical input from the T cell as well as both correct and novel connections from other neurons (Zhang and Nicholls 1983).

5.3.3.3 Mutants Elucidate Synaptic Specificity

The characterization of individual muscles by particular lectins and the generation of antibodies against glycoproteins have begun to elucidate the basis of specificity in arthropod neuromuscular innervation. Another approach is to use mutants to identify molecules that may be expressed for a brief period of time in a small subset of neurons or targets (Reichert 1994). Mutants defective in muscle innervation in *Drosophila* include *clueless* and *walkabout* (Van Vactor et al. 1993). A gene named *passover* is expressed only in descending giant axons and their targets (Krishnan et al. 1993). Like the restricted expression of the gene product of *passover*, the glycoprotein called TERM-1 is localized in axonal endings of two pairs of identified interneurons. It accumulates in discrete sites on the terminals, a situation remindful of the apCAM accumulation at synaptic target sites on motorneuron L7 (Zhu et al. 1994).

The failure of interneurons to recognize the correct synaptic target has been traced to the actions of a single gene in the nematode *Caenorhabditis elegans* (Miller et al. 1992; White et al. 1992). The gene codes for a homeodomain protein. Homeodomain proteins are usually transcription factors, controlling the expression of specific genes. When this one is defective, one class of motoneurons is functionally converted into another class of motoneurons, receiving the interneuronal input appropriate for those cells. This difference occurs in the absence of any alteration in the distribution of processes or the projection pattern of the motoneurons. Identification of the target genes that fail to be expressed should lead to the characterization of features important in specifying synaptogenesis. A gene has also been identified that apparently plays a role in synaptogenesis by *Caenorhabditis* locomotor-circuit neurons, which change synaptic specificities during postembryonic development (Walthall et al. 1993).

5.4 Development of a Functioning Synapse

Once a growth cone encounters an appropriate target, growth will typically cease and a stable presynaptic ending will form. The presynaptic cell must synthesize the transmitter(s) and release vesicle-bound transmitter by a Ca^{2+} dependent mechanism (excitation-secretion coupling). Structural underpinnings of vesicle release in the presynaptic cell involve the accumulation of appropriate Ca^{2+} channels to the vesicle docking sites. The postsynaptic cell must make receptors for the transmitter, localize the receptors at the subsynaptic site, and add any molecular specializations that limit the duration of transmitter effect, such as uptake or digestive enzymes. The conductances that a particular synapse can activate may be regulated in the process of synapse formation. Neurons differ in how ready they are for the process of synaptogenesis, and in how the acquisition of the necessary specializations is regulated.

5.4.1 Neurotransmitters in Synapse Formation

Transmitter synthesis begins in many neurons close to the time of process extension, and differentiated neurons retain this feature. The neurons in the CNS environment interact with growth cones; this non-synaptic function of transmitter release is described in Chapter 4. Developing and regenerating snail and leech neurites experience and respond to neurotransmitters in the environment. Whether the transmitter is released by a potential synaptic partner or not, exposure can have a formative effect. In cultured *Aplysia* sensory neurons involved in the gill withdrawal behavior, bath application of serotonin increases the excitability by depressing the serotonin-sensitive K^+ current, or S-current (Dale et al. 1987). This relatively long-term effect of the transmitter is mediated by protein synthesis. Serotonin also enhances the growth of synaptic varicosities between these sensory neurons and motoneurons (Glanzman et al. 1990).

Neurotransmitter release is not involved in the selection of a muscle target in insects, because release of transmitter does not occur during motoneuron pathfinding. Neuropeptide expression in insects may be initiated by interactions with target cells, as described below for *Drosophila* motoneurons, but other insect neurons express transmitter according to a developmental timetable (Wall and Taghert 1991). The extent to which the release of transmitter by growing insect neurons can influence other neurons is not known; in at least some cases, the synthesis of the postsynaptic receptor is dependent upon synaptic contact.

5.4.2 Targets May Induce Excitation-Secretion Coupling

Prior to formation of a synapse, snail neurons release transmitter spontaneously. An example of rapid synaptogenesis is provided by the connection of *Helisoma* buccal neurons B5 on B19 (Haydon 1988). These neurons are not normally synaptic partners, but they will form a chemical synapse in situ and in vitro under conditions that promote neurite extension. All B5 neurons develop excitatory, ACh-mediated synapses onto B19. Neurons that are cultured separately for 3 days, allowing adjustment to culture conditions, are able to form chemical synapses within seconds of contact (Haydon 1988). In fact, neuron B5 is ready to rapidly form fully functional synapses with all ACh-sensitive targets, including muscle cells that are not normal targets (Zoran et al. 1990). Apparently, this neuron possesses presynaptic secretory function prior to target contact.

In contrast to the example above, formation of synapses by B19 onto dissociated supralateral radular tensor muscle cells follows a timecourse (more than 30 minutes) consistent with activation of protein synthesis. Contact with the target is required to initiate acquisition or activation of excitation-secretion coupling, so that the neuron is converted from one that releases ACh continuously to one in which the release is coupled to cell depolarization (Zoran et al. 1990).

The method of isolation of the muscle cells is critical for the formation of the chemical synapses studied by Zoran et al. (1990). When the isolation protocol includes trypsin, the connections formed between buccal neurons and muscle fibers are exclusively electrical (Cohan et al. 1990). Enzymatic treatment apparently alters surface molecules involved in recognition and initiation of the steps leading to synapse formation.

5.4.3 Alterations in Ca^{2+} Channel Distribution

In synapse formation, the ionic currents may be altered and this may lead to additional steps in neuronal differentiation (Spitzer 1991). Changes in the distribution or function of Ca^{2+} channels as a function of synapse formation have been studied in leech neurons and snail neurons. In the leech Retzius cells, Ca^{2+} currents elicited by depolarization were studied. The formation of a chemical synapse with any of several follower cells is associated with preservation of the Ca^{2+} current levels in the axon stump where the synapse forms, but reduction in the Ca^{2+} responses in the soma (Fernandez-de-Miguel et al. 1992). The formation of electrical synapses does not affect Ca^{2+} channel distribution. The restriction of Ca^{2+} channels to the synaptic region may allow the excitation-secretion events to be focused there. The reduction of distant Ca^{2+} channel must be mediated by second messengers. The Ca^{2+} currents in the postsynaptic cells were also reduced compared to their levels prior to synapse formation, and a similar reduction in voltage-sensitive Ca^{2+} channels is described during synaptogenesis in a snail (Richmond et al. 1991).

Following contact between cultured *Helisoma* buccal motoneurons and appropriate target muscle fibers, action potential-evoked Ca^{2+} accumulation in the presynaptic endings increases almost fourfold in a process dependent upon cyclic AMP-dependent protein kinase activation (Funte and Haydon 1993; Zoran et al. 1993). The internal Ca^{2+} concentration is elevated within 30 min in the region of the synapse, and after 24 h, throughout the cell (Fig. 5.18) (Zoran et al.

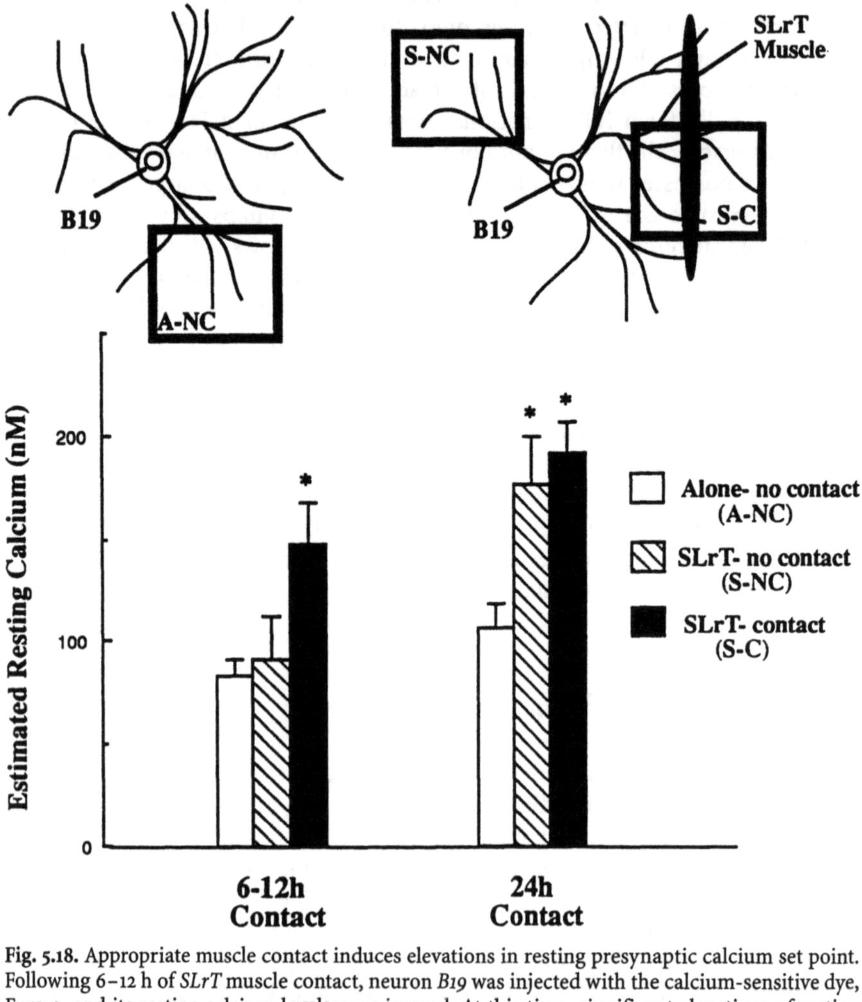

Fig. 5.18. Appropriate muscle contact induces elevations in resting presynaptic calcium set point. Following 6–12 h of *SLrT* muscle contact, neuron *B19* was injected with the calcium-sensitive dye, Fura-2, and its resting calcium levels were imaged. At this time, significant elevations of resting calcium were restricted to neuritic arbors in direct contact *(S-C)* with muscle fibers as compared to neurites of the same neuron which did not directly contact muscle *(S-NC)* and in comparison to the neurites of neurons cultured *alone (A-NC)*. Following 24 h of muscle contact, resting calcium levels in all the neuritic arbors of neuron B19, both in neurites that directly contacted muscle *(S-C)* and in those that had no direct muscle contact *(S-NC)*, are significantly elevated above calcium levels for neurons cultured alone for the same time period. Drawings illustrate typical areas of neuritic arbors which were imaged for calcium measurements. Calcium levels illustrated are means for multiple arbors from many cells ± SEM. (Haydon and Zoran 1994)

1993). The elevation following synaptogenesis is not great enough to halt growth, and may initially function in regulating the cytoskeleton. Its possible relation to vesicle release was not discussed by the authors. A similar elevation of free Ca^{2+} is seen following muscle fiber innervation by *Xenopus* motoneurons (Dai and Peng 1993).

Action-potential-dependent Ca^{2+} accumulation in the buccal neuron is dependent upon maintained synaptic contact with the muscle fiber. The same elevation of resting Ca^{2+} level is also produced by adding a membrane-associated, trypsin-sensitive molecule derived from the muscle membrane to the motoneuron culture (Funte and Haydon 1993).

Although the effect of synaptogenesis on Ca^{2+} was different in these two examples, reduction of Ca^{2+} channels in distant regions of the leech neuron membrane may reduce the tendency for those regions to form synapses after one contact has been made, and elevation of Ca^{2+} within the snail neuron, although not enough to halt growth, may have the effect of allowing shape adjustments to accommodate synapse formation. Changes in internal Ca^{2+} may best be interpreted in the light of the Ca^{2+} set-point hypothesis (Kater and Mills 1991).

5.4.4 Electrical Activity in Synapse Formation

Axons that have navigated a great distance, following chemical cues, can probably be expected to use the same mechanisms to choose a target. Thus it is no surprise that synaptogenesis in many systems can occur in the presence of tetrodotoxin (TTX), the classic blocker of inward current channels of the action potential. TTX in the bathing medium does not alter the formation or breakage of transient electrical synapses in identified, cultured neurons of *Helisoma* (Berdan and Bulloch 1990). Similarly, connections between regenerating cricket cercal afferents and interneurons are not affected by the presence of TTX (Chiba and Murphey 1991).

In contrast to the situation in vertebrate neuromuscular synapse formation, in which secretion of chemical signals (agrin, Campanelli et al. (1992); heregulin ARIA, Falls et al. 1993) is independent of electrical activity, the development of synaptic function in *Drosophila* is dependent upon neural activity (Broadie and Bate 1993a,b). Mutants deficient in ion channel genes reveal that presynaptic electrical activity is required for neural induction of transmitter receptor expression and localization of the receptors at the synaptic site (Broadie and Bate 1993c). Because in denervated muscles the transmitter receptors tend to become diffusely distributed over the muscle membrane, it is reasonable to assume that electrical activity in regenerating motoneurons will similarly be important in attracting receptors to the synaptic site.

5.4.5 Responses to Growth Cone Contact with Target

There is increasing evidence for retrograde signals in the development of transmitter release properties (Davis and Murphey 1994). In leech neurons, restriction of the conductances activated by a synaptic transmitter may precede synapse formation. Retzius (R) cells in the midbody region of the leech respond differently to ACh application, depending on whether or not they innervate the reproductive organs (Kristan et al. 1993). These differences arise during development as a function of interactions with reproductive duct tissue.

Selection of the postsynaptic conductance in the process of synapse formation has been studied in vitro with R and pressure sensitive (P) cells. In the isolated P cell, bath application of serotonin activates two conductances mediated by different receptors: a Cl^- conductance and a monovalent cation conductance (Henderson 1983). The formation of synapses between cultured R and P cells is preceded by a dramatic decline in the monovalent cation conductance of P cells, but the Cl^- conductance is unchanged (Drapeau and Sanchez-Armass 1988, 1989). This effect is localized to the site of the synapse (Fig. 5.19; Ching et al. 1993). The cation channels do not immediately disappear; however, protein kinase C activation, which functions in the receptor cascade, becomes nonfunctional (Drapeau 1990), and kinetics suggest that second messengers alter the sensitivity of the cascade to protein kinase C (Catarsi and Drapeau 1992).

The action of the R cell on the P cell conductance could not be mimicked by bath exposure to serotonin, but could be produced by physical contact between the P cell and a fixed Retzius cell (Drapeau et al. 1989) but not other cells (Merz and Drapeau 1992). This suggests that specific surface molecules are relevant for communication between these synaptic partners.

A different conductance modification is seen in the in vivo regeneration of leech neurons after connective cuts. In this case, the excitatory aspect of the dual sign synapses between pressure and L motor cells is lost (Jansen and Nicholls 1972). It is possible that altered connectivity between regenerated leech neurons and neurons in other animals may represent a failure in the differentiated neurons to recapitulate the steps whereby appropriate synaptic conductances are selected.

In insect motor innervation, activity-independent mechanisms (selective cell adhesion) govern target selection. Synaptic transmission begins within minutes of contact, and both the transmitters and receptors are initially expressed following target contact by filopodia (Broadie and Bate 1993a). As in vertebrate neuromuscular development (Hall and Sanes 1993), the *Drosophila* motoneuron regulates expression of neurotransmitter receptors and their concentration at the synapse.

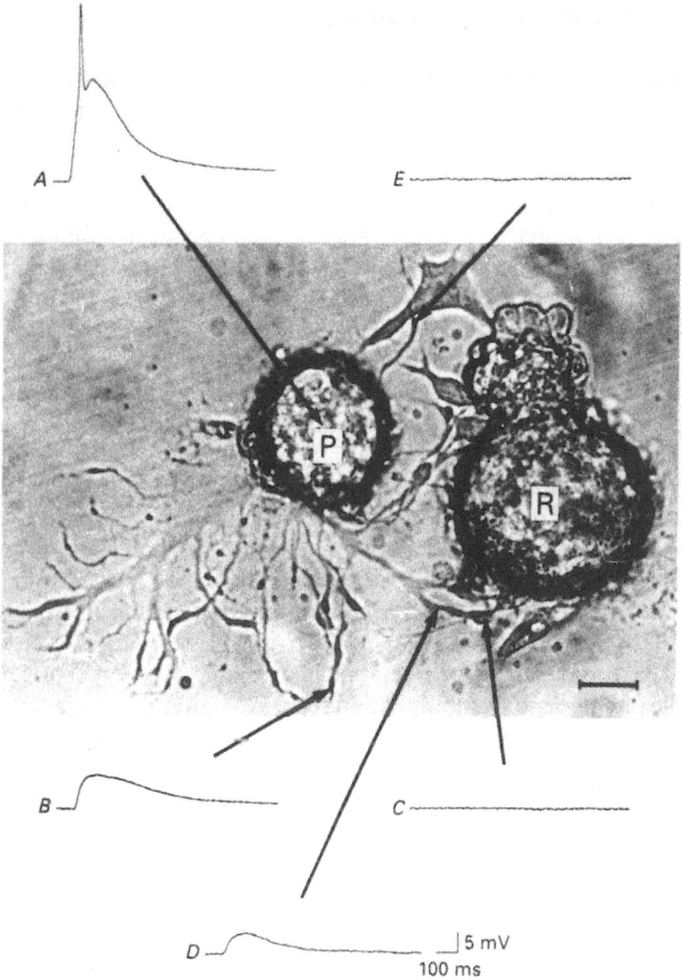

Fig. 5.19. Extrasynaptic response to 5-HT in P cell neurites and growth cones following R-cell contact. The phase-contrast photomicrograph shows an R and P cell pair after 2 days in culture. Intracellular recording with a 4 M cesium acetate-filled electrode measured the responses to focal pressure ejection of 50 mM 5-HT from a micropipette. **A** Application of 5-HT to the P cell soma evoked an action potential. **B** Application of 5-HT to P cell growth cones not contacted by the R cell resulted in a large depolarization. In P cell growth cones contacted by an R cell neurite (**C**) or growth cones **E**, the depolarizing response to 5-HT was absent. **D** Movement of the application pipette 20 μm proximal to the point of contact of the neurite shown in C elicited a depolarizing response. (Ching et al. 1993)

5.5 The Role of Competition

Synaptic competition requires a minimum of three elements: the two input synapses and the postsynaptic cell as the intermediary (Stent 1973; Lichtman and Balice-Gordon 1989). In a competitive situation, if one of the inputs is eliminated or made less competitive, the other is predicted to become stronger. This is because the inputs compete for some resource that is required for expansion. The two branches of a regenerating axon experience differential axonal transport after one branch contacts a target (Goldberg and Schacher 1987).

The role of activity in competition is a complex one. In general, activity in one pathway will make that pathway more competitive than other pathways impinging upon the same target (Fig. 5.17). Whether the other inputs will actually shrink depends upon whether the target has the capacity to expand in proportion to the additional input strength or will maintain a constant total input load (see below).

Synaptic elimination as a result of competition is typically associated with individually powerful inputs, such as motor innervation and climbing-fiber to Purkinje cell inputs (Lichtman and Balice-Gordon 1989; Goodman and Shatz 1993). Competition can space the processes of a single neuron as well as determine the distribution and relative strength of the inputs from different neurons (Macagno et al. 1989).

Dependence upon interactions involving a postsynaptic cell distinguishes the phenomenon of competition from avoidance or mutual inhibition operating among processes of the presynaptic cell population, which can result in a similar segregation of inputs (Macagno et al. 1989). Interactions between homologous neurons rather than competition for a peripheral target determines the morphology and differentiation of leech HA, AE and AP neurons (Gao and Macagno 1987a,b).

A competitive situation may be created by the presence of the postsynaptic cell, as is the case for *Aplysia* sensory neurons and their target cell (Glanzman et al. 1991). Neurites of sensory neurons grown in the absence of a target contact one another freely and grow along common paths, as would be expected for sensory axons growing in a central tract or peripheral nerve. However, when the sensory neurons are cultured with a target neuron (L7), their processes generally segregate, and their contacts are spaced at intervals along the target.

In the *Aplysia* system, the strongest evidence for competition comes from comparing the strength of the inputs. Not only do the sensory neurons divide up the terrain, but they divide up the input carrying capacity of the motoneuron, as measured by EPSP size. The EPSP of a single sensory neuron synapsing on L7 is approximately the same as the sum of the EPSP's of both inputs when two sensory neurons are grown with the motor cell (Glanzman et al. 1991). It is not apparent what the *Aplysia* sensory neurons are competing for, but it is not synaptic space per se, because the target neuron clearly has additional membrane area. Other examples of interactions that adjust the strength of synapses relative to the input carrying capacity of the postsynaptic cell are discussed in section 5.2.4.

The role of competition is apparent when neurons in neighboring ganglia compensate following neuron deletions.

Examples of the regulation of synaptic inputs by competition are seen in the regeneration of the central connections of the leech. After connective cuts, the lateral nociceptive (N) cells normally regenerate their basket-like axosomatic wrappings around cells in adjacent ganglia but they can form substitute synapses with the homologues of their normal target within their own ganglion if isolation blocks access to adjacent ganglia (Gu and Muller 1990). Protease injection into N cells in adjacent ganglia denervates neurons and induces sprouting and wrapping of the cells by intact N cells in the same ganglion. This sprouting and synapse formation cannot be induced by injury to the N cell in the absence of denervation of the target, so competitive interactions apparently regulate access to the target (Gu and Muller 1990). Substitution of homologous targets reveals a hierarchy of acceptable synaptic choices (Muller and Gu 1991).

Increases in synaptic input strength are necessary to maintain a constant level of control over the growing muscle fibers. This is achieved by increases in transmitter release correlated with increases in the number, size and complexity of release sites (Govind 1992). The crustacean neuromuscular synapse adjusts to the altered activity with alterations in synaptic structure and recruitment of silent synapses (Atwood and Wojtowicz 1986; Atwood et al. 1989a; Lnenicka and Atwood 1989; Wojtowicz et al. 1989, 1991; Bittner and Baxter 1991; Lnenicka et al. 1991).

As is true of the changing nature of sensory innervation with growth in arthropods, the adjustment of motor innervation with growth and activity is correlated with its regulation by competition. Competition is a strong force in the maintenance and regulation of motor innervation. Excitatory motor innervation of spiny lobsters is strengthened (a 5-fold increase in quantal content) 1 day after pronase injection kills the other excitatory neuron, and this change precedes anatomical evidence of sprouting (Parnas et al. 1984). Similar adjustment of the level of input strength onto a muscle occurs as a function of activity in one of two excitatory motoneurons in the lobster (Bradacs et al. 1991).

5.6 Conclusions

From study of the effects of injury and ablations on whole animals, we have learned that when outgrowth occurs, correct targets are selected more often than not. However, neural regeneration rarely replicates the original condition. One pattern of less-than-perfect repair includes synaptic connections that are weaker than the original connections. Anatomical correlates of the physiological observations reveal fewer and smaller synaptic connections, diminished synaptic contact area on the postsynaptic cell, and increased responsiveness to other inputs. Such alterations in temporarily interrupted connectivity are the consequence of competitive interactions. Where a major loss of inputs must be accommodated, the increasing contribution of intact inputs may restore function. This is perhaps best illustrated in the recovery of central pattern generators following ablation of sensory structures, but it also applies to motor coordination and central pathways. This suggests that the animal is better off to have competition as a mechanism that appears to compete with the restoration of neural circuits, even if some distortion is the inevitable consequence.

Another way in which regenerated connections may differ from intact circuitry is in the different conductances elicited at a synapse. This situation may result from incompetence of adult neurons to replace the conductance state selected in the original formation of the synapse. The maladaptive consequences of this condition are obvious, and no redeeming features of such a response are apparent. It remains to be seen whether neurons that change transmitter expression as a consequence of hormonal experience at metamorphosis will retain the adult pattern through regeneration.

Most of what we know about the mechanisms of synapse formation by adult neurons has come from in vitro studies. The conductances present in each neuron, the release and responses to neurotransmitters, and the effects of neuromodulators and growth factors on the formation of the synapse can be observed. From these studies, we see the two-way communication between presynaptic and postsynaptic cells: recognition may occur by stages, based upon the degree of membrane overlap and adhesive compatibility. Incorrect, "trial" synapses may occur that allow the neuron to compare the available options before eliminating some in favor of others, according to a preference hierarchy. Alternatively, specific markers on neurons and their targets may allow unerringly correct synapses to be formed with only the input available through filopodial contact. Appropriate contacts activate cascades that may alter the cytoskeleton, the conductances in both cells, the synthesis of synapse-specific molecules, and the distribution of adhesion molecules and receptors.

The steps involved in putting together the synapse are not a pattern activated once during the embryo. The responses initiated when the neuron has contacted its target involve mechanisms continually available to the neurons as they modify and adjust the synapse to match the demands on the circuit. It is the fundamental plasticity of nervous systems that offers great hope for improving the outcome following neural injury. The roadblocks to more effective repair can be dismantled as simple nervous systems reveal to us what the nervous system is capable of accomplishing and how it goes about it.

References

Acklin SE, Nicholls JG (1990) Intrinsic and extrinsic factors influencing properties and growth patterns of identified leech neurons in culture. J Neurosci 10:1082–1090

Aguayo AJ, Rasminsky M, Bray CM, Carbonetto S, McKerracher L, Villegas-Pérez MP, Vidal-Sanz M, Carter DA (1991) Degenerative and regenerative responses of injured neurons in the central nervous system of adult mammals. Philos Trans R Soc Lond Biol Sci 331:337–343

Alberini CM, Ghirardi M, Metz R, Kandel ER (1994) C/EBP is an immediate-early gene required for the consolidation of long-term facilitation in *Aplysia.* Cell 76:1099–1114

Allen RD, Allen NS, Travis JL (1981) Video-enhanced contrast, differential interference contrast (AVEC-DIC) microscopy: a new method capable of analysing microtubule-associated motility in the reticulopodial network of *Allogromia laticollaris.* Cell Motil 1:291–302

Allison P, Benjamin PR (1985) Anatomical studies of central regeneration of an identified molluscan interneuron. Proc R Soc Lond Biol Sci 226:135–157

Aloe L, Bracci-Laudiero L, Alleva E, Lambiase A, Micera A, Tirassa P (1994) Emotional stress induced by parachute jumping enhances blood nerve growth factor levels and the distribution of nerve growth factor receptors in lymphocytes. Proc Natl Acad Sci USA 91: 10440–10444

Altrup U, Kolbe G (1982) Changes in neuronal arborizations induced by lesions of the ganglionic perineurium (*Helix pomatia*). Neurosci Lett 34:215–220

Alvarez-Buylla A, Nottebohm R (1988) Migration of young neurons in adult avian brain. Nature 335:353–354

Ambron RT, Protic J, Den H, Gabel CA (1989) Identification of protein-bound oligosacchairdes on the surface of growth cones that bond muscle cells. J Neurobiol 20: 549–568

Anderson PAV (1989) Evolution of the first nervous systems. Plenum New York, pp 423.

Anderson H, Edwards JS, Palka J (1980) Developmental neurobiology of invertebrates. Annu Rev Neurosci 3:97–139

Anderson PAV, Schwab WE (1982) Recent advances and model systems in Coelenterate neurobiology. Prog Neurobiol 19:213–236

Anderson PAV, Spencer AN (1989) The importance of cnidarian synapses for neurobiology. J Neurobiol 20:435–457

Anderson PAV, Schwab WE, Gilbert S, Allen RD (1986) Fast axonal transport by neurons from the jellyfish *Cyanea capillata.* J Neurobiol 17: 29–37

Arbas EA, Weidner MH (1991) Transneuronal induction of muscle atrophy in grasshoppers. J Neurobiol 22: 536–546

Aréchiga H, Chiquet M, Kuffler D, Nicholls JG (1986) Formation of specific connections in culture by identified leech neurones containing serotonin, acetylcholine and peptide transmitters. J Exp Biol 126:15–31

Atashi JR, Klinz SG, Ingraham CA, Matten WT, Schachner M, Maness PF (1992) Neural cell adhesion molecules modulate tyrosine phosphorylation of tubulin in nerve growth cone membranes. Neuron 8:831–842

Atwood HL, Wojtowicz JM (1986) Short-term and long-term plasticity and physiological differentiation of crustacean motor synapses. Int Rev Neurobiol 28:275–362

Atwood HL, Dixon D, Wojtowicz JM (1989a) Rapid induction of long-lasting synaptic changes at crustacean neuromuscular junctions. J Neurobiol 20:373–385

Atwood HL, Dudel J, Feinstein N, Parnas I (1989b) Long-term survival of decentralized axons and incorporation of satellite cells in motor neurons of rock lobsters. Neurosci Lett 101:121–126

Avel M (1961) L'influence du systeme nerveux sur la regeneration chez les urodeles et les oligochetes. Bull Soc Zool Fr 86: 464–483

Avery L, Horvitz HR (1987) A cell that dies during wild-type *C. elegans* development can function as a neuron in a *ced-3* mutant. Cell 51:1071–1078

Azhderian EM, Hefner D, Lin CH, Kaczmarek LK, Forscher P (1994) Cyclic AMP modulates fast axonal transport in aplysia bag cell neurons by increasing the probability of single organelle movement. Neuron 12:1223–1233

Bacon JP, Murphey RK (1984) Receptive fields of cricket giant interneurons are related to their dendritic structure. J Physiol 352:601–623
Bacon JP, Blagburn JM (1992) Ectopic sensory neurons in mutant cockroaches compete with normal cells for central targets. Development 115: 773–784
Baguna J (1976) Mitosis in the intact and regenerating planarian *Dugesia mediterranea* n. sp. II. Mitotic studies during regeneration and a possible mechanism of blastema formation. J Exp Zool 195:65–80
Baguna J, Ballister R (1978) The nervous system in planarians: peripheral and gastroderman plexuses, pharynx innervation, and the relationship between central nervous system structure and the acoelomate organization. J Morphol 155:237–252
Baguna J, Salo E, Collet J, Auladell C, Ribas M (1988) Cellular, molecular and genetic approaches to regeneration and pattern formation in planarians. Fortschr Zool 36:65–78
Baguna J, Salo E, Auladell C (1989a) Regeneration and pattern formation in planarians III. Evidence that neoblasts are totopotent stem cells and the source of blastema cells. Development 107: 77–86
Baguna J, Salo E, Romero R (1989b) Effects of activators and antagonists of the neuropeptides substance P and substance K on cell division in planarians. Int J Dev Biol 33:261–264
Bähr M, Eschweiler GW (1991) Regenerating adult rat retinal axons reconnect with target neurons in-vitro. Neuroreport 2:581–584
Baier H, Bonhoeffer F (1991) Axon guidance in vitro by a target-derived cell membrane component. In: Letourneau PC, Kater SB, Macagno ER (eds) The nerve growth cone. Raven Press, New York, pp 195–206
Baier H, Bonhoffer R (1992) Axon guidance by gradients of a target-derived component. Science 255:472–476
Baier H, Bonhoeffer R (1994) Attractive axon guidance molecules. Science 265:1541–1542
Bailey CH, Chen M (1989) Structural plasticity at identified synapses during long-term memory in *Aplysia.* J Neurobiol 20:356–372
Bailey CH, Chen M, Keller F, Kandel ER (1992) Serotonin-mediated endocytosis of apCAM: an early step of learning-related synaptic growth in *Aplysia.* Science 256:645–649
Baker MW, Vohra MM, Croll RP (1993) Serotonin depleters, 5,7-dihydroxytryptamine and p-chlorophenylalanine, cause sprouting in the CNS of the adult snail. Brain Res 623: 311–315
Bamburg JR, Bray D, Chapman K (1986) Assembly of microtubules at the tip of growing axons. Nature 321:788–790
Bandtlow CE, Schmidt MF, Hassinger TD, Schwab ME, Kater SB (1993) Role of intracellular calcium in NI-35-evoked collapse of neuronal growth cones. Science 259:80–83
Bannatyne BA, Blackshaw SE, McGregor M (1989) New growth elicited in adult leech mechanosensory neurones by peripheral axon damage. J Exp Biol 143:419–434
Baptista CA, Macagno ER (1988) Modulation of the pattern of synaptic connections of a leech motor neuron by ablation or transplantation of its target. Neuron 1:949–962
Baptista CA, Gershon TR, Macagno ER (1990) Peripheral organs control central neurogenesis in the leech. Nature 346:855–858
Barker DL, Wong RG, Kater SB (1982) Separate factors produced by the CNS of the snail *Helisoma* stimulate neurite outgrowth and choline metabolism in cultured neurons. J Neurosci Res 8:419–432
Bartel DP, Sheng M, Lau LF, Greemberg ME (1989) Growth factors and membrane depolarization activate distinct programs of early response gene expression: dissociation of *fos* and *jun* induction. Genes Dev 3:304–313
Barthalay Y, Hipeau-Jaquotte R, de la Escalera S, Jimenez F, Poivant M (1990) *Drosophila* neurotactin mediates heterophilic cell adhesion. EMBO J 9:3603–3609
Bastiani MJ, Harrelson AL, Snow PM, Goodman, CS (1987) Expression of fasciclin I and II glycoproteins on subsets of axon pathways during neuronal development in the grasshopper. Cell 48:745–755
Bate CM (1976) Pioneer neurons in an insect embryo. Nature 260:54–56
Baylor DA, Nicholls JG (1971) Patterns of regeneration between individual nerve cells in the central nervous system of the leech. Nature 232:268–270

Beadle DJ, G Lees, SB Kater (eds) (1988) Cell culture approaches to invertebrate neuroscience. Academic Press, London, 240 pp
Beck K, Hunter I, Engel J (1990) Structure and function of laminin: anatomy of a multidomain glycoprotein. FASEB J 4:148–160
Bedlack RS Jr, Wei M-d, Loew LM (1992) Localized membrane depolarizations and localized calcium influx during electric field-guided neurite growth. Neuron 9:393–403
Belardetti F, Schacher S, Kandel ER, Siegelbaum SA (1987) Neuronal inhibition by the peptide FMRFamide involves opening of S K channels. Nature 325: 153–156
Benbassat D, Spira ME (1993) Survival of isolated axonal segments in culture: morphological, ultrastructural, and physiological analysis. Exp Neurol 122: 295–310
Benbassat D, Spira ME (1994) The survival of transected axonal segments of cultured *Aplysia* neurons is prolonged by contact with intact nerve cells. Eur J Neurosci 6:1605–1614
Benjamin PR, Allison P (1985) Regeneration of excitatory, inhibitory and biphasic synaptic connections made by a snail giant interneuron. Proc R Soc Lond B Biol Sci 226:159–176
Benjamin PR, Allison P (1987) Regeneration of a molluscan multiaction interneurone: a cellular model for the control of chemical synapse regeneration and sprouting. In: Winlow W, McCrohan CR (eds) Growth and plasticity of neural connections, Manchester Univ Press, Manchester, pp 57–79
Bentley D, Toroian-Raymond A (1986) Disoriented pathfinding by pioneer neuron growth cones deprived of filopodia by cytochalasin treatment. Nature 323: 712–715
Bentley D, Guthrie PB, Kater SB (1991) Calcium ion distribution in nascent pioneer axons and coupled preaxogenesis neurons in situ. J Neurosci 11:1300–1308
Berdan RC, Bulloch AGM (1990) The role of activity in the selection of new electrical synapses between adult *Helisoma* neurons. Brain Res 206:129–144
Berdan RC, Easaw JC (1992) Modulation of sprouting in organ culture after axotomy of an identified molluscan neuron. J Neurobiol 23:433–450
Berdan RC, Ridgway RL (1992) Release of neurite outgrowth promoting factors by *Helisoma* central ganglia depends on neural activity. Brain Res 572:132–138
Berdan RC, Jones PG, Bulloch AGM (1989) Modulation of neurite outgrowth: ultrastructure and physiology. In: Bulloch AGM (ed) The cellular basis of neuronal plasticity, Manchester Univ Press, Manchester, pp 97–128
Berdan RC, Hauser G, Bulloch AGM (1990) Ultrastructure of an identified molluscan neuron in organ culture and cell culture following axotomy. J Comp Neurol 296:437–446
Berdan RC, Easaw JC, Wang R (1993) Alterations in membrane potential after axotomy at different distances from the soma of an identified neuron and the effect of depolarization on neurite outgrowth and calcium channel expression. J Neurophysiol 69:151–164
Berrill NJ (1961) Growth, development and pattern. WH Freeman, San Francisco pp555
Bickell-Page LR, Mackie GO (1991) Tentacle autotomy in the hydromedusa *Aglantha digitale* (Cnidaria): an ultrastructural and neurophysiological analysis. Philos Trans R Soc Lond B Biol Sci 331:155–170
Birse SC, Bittner GD (1976) Regeneration of giant axons in earthworms. Brain Res 113:575–581
Bittner GD (1977) Trophic interactions of crustacean neurons In: G Hoyle (ed) Identified neurons and behavior. Plenum Press, New York, pp 507–532
Bittner GD (1988) Long term survival of severed distal axonal stumps in vertebrates and invertebrates. Am Zool 28:1165–1179
Bittner GD (1991) Long-term survival of anucleate axons and its implications for nerve regeneration. Trends Neurosci 14: 183–188
Bittner GD, Baxter DA (1991) Synaptic plasticity at crayfish neuromuscular junctions: facilitation and augmentation. Synapse 7:235–243
Bittner GD, Johnson AL (1974) Degeneration and regeneration in crustacean peripheral nerves. J Comp Physiol 89: 1–21
Bittner GD, Ballinger ML, Larimer J (1974) Crayfish CNS: minimal degenerative regenerative changes after lesioning. J Exp Zool 180:13–36
Bixby JL (1989) Protein kinase C is involved in laminin stimulation of neurite outgrowth. Neuron 3:287–297

Blackshaw SE, Nicholls J, Parnas I (1982) Expanded receptive fields of cutaneous mechanoreceptor cells after single neurone deletion in leech central nervous system. J Physiol Lond 326:251–260
Blair SS, Murray MA, Palka J (1987) The guidance of axons from transplanted neurons through aneural *Drosophila* wings. J Neurosci 7:4165–4175
Blanco RE (1990) Ultrastructural changes in glial cells during regeneration of cockroach peripheral nerve. Glia 3:98–109
Blanco RE, Lane NJ (1990) Changes in intercellular junctions during peripheral nerve regeneration in insects. J Neurocytol 19:873–882
Bliss DE (1960) Autotomy and regeneration. In: Waterman TH (ed) The physiology of crustacea, vol 1, Academic Press, New York, pp 561–589
Blundon JA, Sheller RA, Moehlenbruck JW, Bittner GD (1990) Effect of temperature on long term survival of anucleate giant axons in crayfish and goldfish. J Comp Neurol 299:377–391
Bode HR (1992) Continuous conversion of neuron phenotype in hydra. Trends Genet 8:279–284
Bode HR, Gee LW, Chow MA (1990) Neuron differentiation in hydra involves dividing intermediates. Dev Biol 139:231–243
Bode HR, Heimfield S, Koizumi O, Littlefield CL, Yaross MS (1988) Maintenance and regeneration in the nerve net of Hydra. Am Zool 28:1053–1063
Bodenstein D (1957) The regeneration of nerves in insects. J Exp Biol 54:443–452
Bodmer R, Dagan D, Levitan IB (1984) Chemical and electrotonic connections between *Aplysia* neurons in primary culture. J Neurosci 4:228–233
Bollner T (1989) Regeneration of the central nervous system of the ascician *Ciona intestinalis.* In: Kiortsis V, Koussoulakos S, Wallace H (eds) Recent trends in regeneration research, NATO Series A; Life Sci 172. Plenum Press, New York, pp 281–287
Bollner T, Beesley PW, Thorndyke MC (1992) Pattern of substance P- and cholecystokinin-like immunoreactivity during regeneration of the neural complex in the ascidian *Ciona intestinalis.* J Comp Neurol 325:572–580
Bollner T, Beesley PW, Thorndyke MC (1993) Distribution of GABA-like immunoreactivity during post-metamorphic development and regeneration of the central nervous system in the ascidian *Ciona intestinalis.* Cell Tissue Res 272:553–561
Bollner T, Howalt S, Thorndyke MC, Beesley PW (1995) Regeneration and post-metamorphic development of the central nervous system in the protochordate *Ciona intestinalis*: a study with monoclonal antibodies. Cell Tissue Res 279: 421–432
Bouton MS, Bittner GD (1981) Regeneration of motor axons in crayfish limbs: distal stump activation followed by synaptic reformation. Cell Tissue Res 219:379–392
Bovolenta P, Wandosell F, Nieto-Sampedro M (1991) Central neurite outrgrowth over glial scar tissue in vitro. In: Letourneau PC, Kater SB, Macagno ER (eds) The nerve growth cone. Raven Press, New York, pp 477–488
Bowling D, Nicholls J, Parnas I (1978) Destruction of a single cell in the nervous system of the leech as a means of analysing its connexions and functional role. J Physiol 282:169–180
Bradacs H, Mercier AJ, Atwood HL (1991) Activity-dependent adaptation of lobster motor neurons and compensation of transmitter release by synergistic inputs. Ann New York Acad Sci 627:378–380
Brandstätter JH, Shaw SR, Meinertzhagen IA (1992) Reactive synaptogenesis following degeneration of photoreceptor terminals in the fly's optic lobe: a quantitative electron microscopic study. Proc R Soc Lond B Biol Sci 247:1–7
Braun J, Stent GS (1989) Axon outgrowth along segmental nerves in the leech. Dev Biol 132:486–501
Bray D (1991) Cytoskeletal basis of nerve axon growth. In: Letourneau PC, Kater SB, Macagno ER (eds) The growth cone. Raven Press, New York, pp 7–17
Briedbach O (1988) Reorganization of identified neurons during the metamorphosis of the meal beetle – aspects for an analysis of neuronal regeneration. Monogr Dev Biol 21: 105–111
Breidbach O, Dennis R, Marx J, Gorlach C, Wiegandt H, Wegerhoff R (1992) Insect glial cells show differential expression of a glycolipid-derived, glucuronic acid-containing epitope throughout neurogenesis: detection during postembryogenesis and regeneration in the

central nervous system of *Tenebrio molitor* L. Neurosci Lett 147:5–8
Broadie K, Bate M (1993a) Development of the embryonic neuromuscular synapse of *Drosophila melanogaster.* J Neurosci 13:144–166
Broadie K, Bate M (1993b) Activity-dependent development of the neuromuscular synapse during *Drosophila* embryogenesis. 11:607–619
Broadie K, Bate M (1993c) Development of larval muscle properties in the embryonic myotubes of *Drosophila melanogaster.* J Neurosci 13:167–180
Brondsted HV (1969) Planarian regeneration. Pergamon Press, Oxford
Brown MC, Holland RL, Hopkins WG (1981) Motor nerve sprouting. Annu Rev Neurosci 4:17–42
Brusca RS, Brusca GJ (1990) Invertebrates. Sinauer Assoc, Sunderland, Massachusetts
Budnik V, Wu C, White K (1989) Altered branching of serotonin-containing neurons in *Drosophila* mutants unable to synthesize serotonin and dopamine. J Neurosci 9:2866–2877
Bulloch AGM (1984) Sprouting and retraction of neurites by undamaged adult molluscan neurons. Brain Res 321: 369–371
Bulloch AGM (1985) Development and plasticity of the molluscan nervous system. In: Willows AOD (ed) The Mollusca, Vol 8. Neurobiology and behavior, part I. Academic Press, Orlando, pp 335–400
Bulloch AGM (1987) Somatostatin enhances neurite outgrowth and electrical coupling of regenerating neurons in *Helisoma.* Brain Res 412:6–17
Bulloch AGM (1989) Sprouting by intact *Helisoma* neurons: role for glutamate. J Neurosci Res 23:384–395
Bulloch AGM, Hauser GC (1990) Sprouting by isolated *Helisoma* neurons: enhancement by glutamate. Int J Dev Neurosci 8:391–398
Bulloch AGM, Kater SB (1981) Selection of a novel connection by adult molluscan neurons. Science 212:79–81
Bulloch AGM, Kater SB (1982) Neurite outgrowth and selection of new electical connections by adult *Helisoma* neurons. J Neurophysiol 48:569–583
Bulloch AGM, Ridgway RL (1989) Neuronal plasticity in the adult invertebrate nervous system. J Neurobiol 20:295–311
Bulloch AGM, Kater SB, Murphy AD (1980) Connectivity changes in an isolated molluscan neuron during in vivo culture. J Neurobiol 11:531–546
Bulloch AGM, Kater SB, Miller HR (1984) Stability of new electrical connections between adult *Helisoma* neurons is influenced by preexisting neuronal interactions. J Neurophysiol 52:1094–1105
Bullock TH, Horridge GA (1965) Part IV. The mollusca. In: Structure and function in the nervous system of invertebrates, vol II. WH Freeman, San Francisco, pp 1273–1515
Büschges A, Pearson KG (1991) Adaptive modification in the flight system of the locust after removal of wing proprioceptors. J Exp Biol 157:313–333
Büschges A, Ramirez J-M, Pearson KG (1992a) Reorganization of sensory regulation of locust flight after partial deafferentation. J Neurobiol 23:31–43
Büschges A, Ramiriz J-M, Driesang R, Pearson KG (1992b) Connections of the forewing tegulae in the locust flight system and their modification following partial deafferentation. J Neurobiol 23:44–60
Byers MR (1970) Chromatolysis in a pair of identified metathoracic neurons in the cockroach, *Diploptera punctata.* Tissue Cell 2:255–279
Cagan R, Ready D (1989) The emergence of order in the *Drosophila* pupal retina. Dev Biol 136:346–362
Cagan RL, Zipursky SL (1992) Cell choice and patterning in the *Drosophila* retina. In: Shankland M, Macagno ER (eds) Determinants of neuronal identity, Academic Press, San Diego, pp 189–224
Camhi JM, Macagno E (1991) Using fluorescence photoablation to study the regeneration of singly cut leech axons. J Neurobiol 22:116–129
Campanelli JT, Ferns M, Hoch W, Rupp F, Von Zastrow M, Hall Z, Scheller RH (1992) Agrin, a synaptic basal lamina protein that regulates development of the neuromuscular junction. Cold Spring Harbor Symp Quant Biol 57:461–472

Cancalon P (1985) Influence of temperature on various mechanisms associated with neuronal growth and nerve regeneration. Prog Neurobiol 25:27–92
Candia-Carnevali MD, Bruno L, Donini SD, Melone G (1989) Regeneration and morphogenesis in the feather star arm. In: Kiortsis V, Koussoulakos S, Wallace H (eds) Recent trends in regeneration research. NATO Series A, Life Sci 172. Plenum Press, New York, pp 447–460
Carbonetto S, Muller KJ (1977) A regenerating neurone in the leech can form an electrical synapse on its severed axon segment. Nature 267:450–452
Carbonetto S, Muller KJ (1982) Nerve fiber growth and the cellular response to axotomy. Curr Top Dev Biol 17:331–376
Caroni P, Schwab, ME (1988) Two membrane protein fractions from rat central myelin with inhibitory properties for neurite growth and fibroblast spreading. J Cell Biol 106:1281–1288
Carroll DJ, Kempf SC (1994) Changes occur in the central nervous system of the nudibranch *Berghia verrucicornis* (Mollusca, Opisthobranchia) during metamorphosis. Biol Bull 186:202–212
Carrow GM, Levitan IB (1989) Selective formation and modulation of electrical synapses between cultured *Aplysia* neurons. J Neurosci 9:3657–3664
Casagrand JL, Ritzman RE (1991) Localization of ventral giant interneuron connections to the ventral median branch of thoracic interneurons in the cockroach. J Neurobiol 22:643–658
Cash D, Carew TJ (1989) A quantitative analysis of the development of the central nervous system in juvenile *Aplysia californica.* J Neurobiol 20:25–47
Cash S, Chiba A, Keshinian H (1992) Alternate neuromuscular target selection following loss of single muscle fibers in *Drosophila.* J Neurosci 12:2051–2064
Catarsi S, Drapeau P (1992) Loss of extrasynaptic channel modulation by protein kinase C underlies the selection of serotonin responses in an identified leech neuron. Neuron 8:275–281
Caudy M, Bentley D (1986) Pioneer growth cone morphologies reveal proximal increases in substrate affinity within leg segments of grasshopper embryos. J Neurosci 6:364–379
Cayre M, Strambi C, Strambi A (1994) Neurogenesis in an adult insect brain and its hormonal control. Nature 368:57–59
Cervero F, Laird JM (1991) Signalling of a step-like intensity change of noxious and mechanical stimuli by dorsal horn neurones in the rat spinal cord. J Physiol (Lond) 434: 561–575
Chalfie M, Sulston JE, White JG, Southgate E, Thomson JN, Brenner S (1985) The neural circuitry for touch sensitivity in *Caenorhabditis elegans.* J Neurosci 5:956–964
Chamak B, Prochiantz A (1989) Influence of extracellular matrix proteins on the expression of neuronal polarity. Development 106:483–491
Chang WS, Serikawa K, Allen K, Bentley D (1992) Disruption of pioneer growth cone guidance in vivo by removal of glycosyl-phosphatidylinositol-anchored cell surface protiens. Development 114:507–519
Chase R (1986) Lessons from snail tentacles. Chem Sens 11:411–426
Chase R, Kamil R (1983) Morphology and odor sensitivity of regenerated snail tentacles. J Neurobiol 14:43–50
Chase R, Rieling J (1986) Autoradiographic evidence for receptor cell renewal in the olfactory epithelium of a snail. Brain Res 384:232–239
Chase R, Tolloczko B (1987) Evidence for differential DNA endoreplication during the development of a molluscan brain. J Neurobiol 18:395–406
Chen JS, Levi-Montalcini R (1970) Axonal growth from insect neurones in glia-free cultures. Proc Natl Acad Sci USA 66:32–39
Chiasson BJ, Baker MW, Croll RP (1994) Morphological changes and functional recovery following axotomy of a serotonergic cerebrobuccal neurone in the land snail A*chatina fulica.* J Exp Biol 192:147–167
Chiba A, Murphey RK (1991) Connectivity of identified central synapses in the cricket is normal following regeneration and blockade of presynaptic activity. J Neurobiol 22:130–142
Chiba A, Shepherd D, Murphey RK (1988) Synaptic rearrangement during postembryonic development in the cricket. Science 240: 901–905
Chiba A, Hing H, Cash S, Keshishian H (1993) Growth cone choices of *Drosophila* motoneurons in response to muscle fiber mismatch. J Neurosci 13:714–732

Chien CB, Rosenthal DE, Harris WA, Holt CE (1993) Navigational errors made by growth cones without filopodia in the embryonic *Xenopus* brain. Neuron 11:237–251
Ching S, Catarsi S, Drapeau P (1993) Selection of transmitter responses at sites of neurite contact during synapse formation between identified leech neurons. J Physiol 468: 425–439
Chiquet M, Acklin SE (1986) Attachment to Con A or extracellular matrix initiates rapid sprouting by cultured leech neurons. Proc Natl Acad Sci USA 83:6188–6192
Chiquet M, Nicholls JG (1987) Neurite outgrowth and synapse formation by identified leech neurones in culture. J Exp Biol 132:191–206
Chiquet M, Masuda-Nakagawa L, Beck K (1988) Attachment to an endogenous laminin-like protein initiates sprouting by leech neurons. J Cell Biol 107:1189–1198
Chitnis AB, Kuwada JY (1991) Elimination of a brain tract increases errors in pathfinding by follower growth cones in the zebrafish embryo. Neuron 7: 277–285
Clark RB, Clark ME, Ruston JG (1962) The endocrinology of regeneration in some errant polychaetes. In: Heller H, Clark RB (eds) Memoires of the society for endocrinology #12. Neurosecretion. Academic Press, New York, pp 275–286
Clement JF, Taylor AK, Vélez SJ (1983) Effect of a limited target area on regeneration of specific neuromuscular connections in the crayfish. J Neurophysiol 49:216–226
Cohan CS (1990) Frequency-dependent and cell-specific effects of electrical activity on growth cone movements of cultured *Helisoma* neurons. J Neurobiol 21: 400–413
Cohan CS, Kater SB (1986) Suppression of neurite elongation and growth cone motility by electrical activity. Science 232: 1638–1640
Cohan CS, Haydon PG, Kater SB (1985) Single channel activity differs in growing and nongrowing growth cones of isolated identified neurons of *Helisoma*. J Neurosci Res 13:285–300
Cohan CS, Conner JA, Kater SB (1987a) Electrically and chemically mediated increases in intracellular calcium in neuronal growth cones. J Neurosci 7:3588–3599
Cohan CS, Haydon PG, Mercier AJ, Kater SB (1987b) Formation, maintenance, and functional uncoupling of connections between identified *Helisoma* neurons in situ. J Neurobiol 18:329–341
Cohan CS, Errick JE, Xia M-H (1990) Formation of electrical connections between cultured identified neurons and muscle fibers in the snail *Helisoma*. Dev Brain Res 56:87–97
Cohen MJ, Jacklet JW (1965) Neurons of insects: RNA changes during injury and regeneration. Science 148:1237–1249
Cohen AH, Baker MT (1988) Functional and non-functional regeneration in the spinal cord of adult lampreys. In: Reier PJ, Bunge RP, Seil FJ (eds) Current issues in neural regeneration research. Alan R Liss, New York, pp 387–396
Cole RN, Morell RJ, Zipser B (1989) Glial processes, identified through their glial specific 130 kD surface glycoprotein, are juxtaposed to sites of neurogenesis in the leech germinal plate. Glia 2: 446–457
Collett TS (1994) Bees learn how to look. Cur Bio 4:717–719
Condic ML, Bentley D (1989a) Removal of the basal lamina in vivo reveals growth cone-basal lamina adhesive interactions and axonal tension in grasshopper embryos. J Neurosci 9:2678–2686
Condic ML, Bentley D (1989b) Pioneer growth cone adhesion in vivo to boundary cells and neurons after enzymatic removal of basal lamina in grasshopper embryos. J Neurosci 9:2687–2696
Condic ML, Bentley D (1989c) Pioneer neuron pathfinding from normal and ectopic locations in vivo after removal of the basal lamina. Neuron 3:427–439
Conner EA, Smith MA (1994) Retrograde signaling in the formation and maintenance of the neuromuscular junction. J Neurobiol 25: 722–739
Constantine-Paton M, Cline HT, Debski, E (1990) Patterned activity, synaptic convergence, and the NMDA receptor in developing visual pathways. Annu Rev Neurosci 13:129–154
Croll RP, Baker MW (1990) Axonal regeneration and sprouting following injury to the cerebral-buccal connective in the snail *Achatina fulica*. J Comp Neurol 300:273–286
Croll RP, Chiasson BJ (1989) Postembryonic development of serotonin-like immunoreactivity in the central nervous system of the snail *Lymnaea stagnalis*. J Comp Neurol 280:122–142

Curran T, Morgan JI (1995) Fos: An immediate-early transcription factor in neurons. J Neurobiol 26:403–412
Czeh G, Gallego R, Kuno M, Kudo M (1978) Evidence for maintenance of motoneurone properties by muscle activity. J Physiol Lond 281:239–252
Dai Z, Peng HB (1993) Elevation in presynaptic Ca^{2+} level accompanying initial nerve-muscle contact in tissue culture. Neuron 10: 827–837
Dale N, Kandel ER, Schacher S (1987) Serotonin produces long-term changes in the excitability of *Aplysia* sensory neurons in culture that depend upon new protein synthesis. J Neurosci 7:2232–2238
Dash PK, Kari KA, Colicos MA, Prywes R, Kandel ER (1991) cAMP response element-binding protein is activated by Ca^{2+}/calmodulin -- as well as cAMP-dependent protein kinase. Proc Natl Acad Sci 88:5061–5065
Davenport RW, Kater SB (1992) Local increases in intracellular calcium elicit local filopodial responses in *Helisoma* neuronal growth cones. Neuron 9:405–416
David CN, Gierer A (1974) Cell cycle kinetics and development of *Hydra attenuata*. III. Nerve and nematocyte differentiation. J Cell Sci 16:359–375
David CN, Hager G (1994) Formation of a primitive nervous system: nerve cell differentiation in the polyp *Hydra*. Perspect Dev Neurobiol 2: 135–140
Davies L, Keenan CL, Koopowitz H (1985) Nerve repair and behavioral recovery following brain transplantation in *Notoplana acticola*. J Exp Zool 235:157–173
Davis GW, Murphey RK (1994) Retrograde signaling and the development of transmitter release properties in the invertebrate nervous system. J Neurobiol 25:740–756
Davis L, Dou P, DeWit M, Kater SB (1992) Protein synthesis within neuronal growth cones. J Neurosci 12:4867–4877
Davis RL (1989) Voltage-dependent properties of electrical synapses formed between identified leech neurones in vitro. J Physiol 417: 25–46
Delaney KR, Zucker RS (1990) Calcium released by photolysis of DM-nitrophen stimulates transmitter release at squid giant synapse. J Physiol Lond 426:473–498
Denburg JL (1982) Elimination of inappropriate axonal branches of regenerating cockroach motor neurons as detected by the retrograde transport of horseradish peroxidase conjugated wheat germ agglutinin. Brain Res 248:1–8
Denburg JL (1985) Plasticity in the cockroach neuromuscular system. Dev Biol 111:306–315
Denburg JL (1988) Cell-cell recognition in the regenerating neuromuscular system of the cockroach. Am Zool 28:1135–1144
Denburg JL (1989) The reappearance of a developmental stage-specific antigen in adult regenerating neurons of the cockroach. J Neurosci 9:3491–3504
Denburg JL, Caldwell RT (1992) A morphological correlate of target recognition by regenerating motor axons in the cockroach. J Comp Neurol 315:364–374
Denburg JL, Hood NA (1977) Protein synthesis in regenerating motor neurons in the cockroach. Brain Res 125:227–239
Denburg JL, Norbeck BA (1989) An axon growth associated antigen is also an early marker of neuronal determination. Dev Biol 135:99–110
Denburg JL, Whitington PM (1978) Regeneration of entire legs in cockroaches as a model for developmental events. Experientia 34 (Separatum): 252–253
Denburg JL, Seecof RL, Horridge GA (1977) The path and rate of growth of regenerating motor neurons in the cockroach. Brain Res 125:213–226
Denburg JL, Eastburn JL, Caldwell RT (1983a) Lectin receptors in the cockroach neuromuscular system. I. Distribution among the set of coxal depressor muscles. Brain Res 289:241–252
Denburg JL, Eastburn JL, Caldwell RT (1983b) Lectin receptors in the cockroach neuromuscular system. II. Effects of denervation on muscle lectin receptors. Brain Res 289:253–267
Denburg JL, Caldwell RT, Marner JM (1987) Differences in surface molecules of motor axon terminals correlated with cell-cell recognition. J Neurobiol 18:407–416
Denburg JL, Powell SL, Murphy BF Jr (1988) Absence of competitive interactions among axon terminals of regenerating motor neurons. J Neurobiol 19:656–665
Denburg JL, Norbeck BA, Caldwell RT, Marner JAM (1989) Developmental stage-specific antigens in the nervous system of the cockroach. Dev Biol 132:1–13

DeRiemer SA, Elliott EJ, Macagno ER, Muller KJ (1983) morphological evidence that regenerating axons can fuse with severed axon segments. Brain Res 272:157–161
Desarmenien GM, Spitzer NC (1991) Role of calcium and protein kinase C in development of the delayed rectifier potassium current in *Xenopus* spinal neurons. Neuron 7: 797–805
Doherty P, Ashton SV, Moore SE, Walsh FS (1991) Morphoregulatory activities of NCAM and N-cadherin can be accounted for by G protein-dependent activation of L-type and N-type neuronal Ca^{2+} channels. Cell 67:21–33
Dotti CG, Sullivan CA, Banker GA (1988) The establishment of polarity by hippocampal neurons in culture. J Neurosci 8:1454–1468
Drapeau P (1990) Loss of channel modulation by transmitter and protein kinase C during innervation of an identified leech neuron. Neuron 4:875–882
Drapeau P, Sanchez-Armass S (1988) Selection of postsynaptic serotonin receptors during reinnervation of an identified leech neuron in culture. J Neurosci 8:4718–4727
Drapeau P, Sanchez-Armass S (1989) Parallel processing and selection of the respnses to serotonin during reinnervation of an identified leech neuron. J Neurobiol 20:312–325
Drapeau P, Melinyshyn E, Sanchez-Armass S (1989) Contact-mediated loss of the non-synaptic response to transmitter during reinnervation of an identified leech neuron in culture. J Neurosci 9:2502–2508
Drewes CD, Fourtner CR (1990) Morphallaxis in an aquatic oligochaete *Lumbriculus*: reorganization of escape reflexes in regenerating body fragments. Dev Biol 138:94–103
Drewes CD, Vining EP, Zoran MJ (1988) Regeneration of rapid escape reflex pathways in earthworms. Am Zool 28:1077–1089
Ebberink RHM, Smit AB, van Minnen J (1989) The insulin family: evolution of structure and function in vertebrates and invertebrates. Biol Bull 177:176–182
Edwards JS (1988) Sensory regeneration in arthropods: implications of homoeosis and of ectopic sensilla. Am Zool 28: 1155–1164
Edwards JS, Palka J (1971) Neural regeneration: delayed formation of central contacts by insect sensory cells. Science 172:591–594
Edwards JS, Palka J (1974) The cerci and abdominal giant fibres of the house cricket, *Acheta domesticus*: I. anatomy and physiology of normal adults. Proc Soc Lond B Biol Sci 185:83–103
Edwards JS, Reddy GR, Rani MU (1989) Central projections of a homoeotic regenerate antennapedia in stick insect *Carausius morosus* (phasmida) J Neurobiol 20:101–114
Egid K, Lnenicka GA (1993) Regeneration from crayfish phasic and tonic motor axons in vitro. J Neurobiol 24:1111–1120
Egid K, Lnenicka GA (1994) Differential effects of depolarization on the survival and growth of two populations of motoneurons in culture. Soc Neurosci Abstr 20:659
Eichler VB (1972) Regeneration of lower vertebrates and invertebrates, vol III. MSS Information Corp, New York
Eisen JS, Myers PZ, Westerfield M (1986) Pathway selection by growth cones of identified neurons in live zebra fish embryos. Nature 320:269–271
Elliott EJ, Muller KJ (1982) Synapses between neurons regenerate accurately after destruction of ensheathing glial cells in the leech. Science 215:1260–1262
Elliot EJ, Muller KJ (1983) Sprouting and regeneration of sensory axons after destruction of ensheathing glial cells in the leech central nervous system. J Neurosci 3:1994–2006
Elphick MR, Emson RH, Thorndyke MC (1989) FMRFamide-like Immunoreactivity in the Nervous System of the Starfish *Asterias rubens*. Biol Bull 177:141–145
Elvin M, Koopowitz H (1994) Neuroanatomy of the rhabdocoel flatworm *Mesostoma ehrenbergii* (Focke, 1836). J Comp Neurol 343:319–331
Ely P, Vélez SJ (1982) Regeneration of specific neuromuscular connections in the crayfish. I. Pattern of connections and synaptic strength. J Neurophysiol 47:656–665
Emson RH, Wilkie IC (1980) Fission and autotomy in echinoderms. Oceanogr Mar Biol Annu Rev 18: 155–250
Faisst J, Keenan CL, Koopowitz H (1980) Neuronal repair and avoidance behavior in the flatworm, *Notoplana acticola*. J Neurobiol 5:483–496
Falls DL, Rosen KM, Corfas G, Lane WS, Fischbach GD (1993) ARIA a protein that stimulates acetylcholine receptor synthesis is a member of the *neu* ligand family. Cell 72:801–815

Fawcett JW (1992) Intrinsic neuronal determinants of regeneration. Trends Neurosci 15:5–8
Fazeli MS (1992) Synaptic plasticity: on the trail of the retrograde messenger. Trends Neurosci 4:115–117
Fekete DM, Brockes JP (1987) A monoclonal antibody detects a difference in the cellular composition of developing and regenerating limbs of newts. Development 99:589–602
Fekete DM, Ferretti P, Gordon H, Brockes JP (1988) Ontogeny of the nerve dependence of amphibian limb regeneration. In: Easter SS, Barald KF, Carlson BM (eds) From message to mind. Sinauer, Sunderland, Massachusetts, pp 29–42
Fenger U, Hofmann M, Galliot B, Schaller HC (1994) The role of the cAMP pathway in mediating the effect of head activator on nerve-cell determination and differentiation in *Hydra*. Mech Dev 47:115–125
Fernandez AD, Flores DV (1978) Studies in the regeneration of the sensory tentacular organs after denervation in *Helix aspersa*. Rev de Microsc Electron 5:186–187
Fernandez-de-Miguel F, Cooper RL, Adams WB (1992) Synaptogenesis and calcium current distribution in cultured leech neurons. Proc R Soc Lond Biol Sci 247: 215–221
Fessler LI, Campbell AG, Duncan KG, Fessler JH (1987) *Drosophila* laminin: characterization and localization. J Cell Biol 105:2383–2391
Fessler JH, Fessler LI (1989) *Drosophila* extracellular matrix. Annu Rev Cell Biol 5:309–339
Fields RD, Neale EA, Nelson PG (1990) Effects of patterned electrical activity on neurite outgrowth from mouse sensory neurons. J Neurosci 10: 2950–2964
Fishman HM, Tewari KP, Stein PG (1990) Injury-induced vesciculation and membrane redistribution in squid giant axon. Biochem Biophys Acta 1023:421–435
Flores V, de Iraldi AP (1973) On the regeneration of the eye in *Helix aspersa* and *Cryptomphalus aspersa*. Z Zellforsch 142: 63–68
Flores V, Brusco A, Scicolone G, Saavedra JP (1992) Serotonergic reinnervation of regenerating tentacular sensory organs in a pulmonate snail, *Cryptomphalus aspersa*. Int J Dev Neurosci 10: 331–340
Forscher P, Kaczmarek LK, Buchanan J, Smith SJ (1987) Cyclic AMP induces changes in distribution and transport of organelles within growth cones of *Aplysia* bag cell neurons. J Neurosci 7:3600–3611
Forscher P, Smith SJ (1988) Actions of cytochalasins on the organization of actin filaments in a neuronal growth cone. J Cell Biol 107:1505–1515
Frank E, Jansen JKS, Rinvik E (1975) A multisomatic axon in the central nervous system of the leech. J Comp Neurol 159: 1–14
Franquinet R (1979) Role de la serotonine et des catecholamines dans la regeneration de la planaire *Polycelis tenuis*. J Embryol Exp Morphol 51: 85–95
Fredman SM, Nutz PG (1988) Regeneration of identified neurons and their synaptic connections in the central nervous system of *Aplysia*. Am Zool 28:1099–1108
French KA, Kristan WB (1994) Cell-cell interactions that modulate neuronal development in the leech. J Neurobiol 25:640–651
French KA, Jordan SM, Loer CM, Kristan WB Jr (1992) Mesenchyme of embryonic reproductive ducts directs process outgrowth of Retzius neurons in the medicinal leech. Dev Biol 153:122–140
Friedlander DR, Levinthal C (1982) Anomalous anatomy of identified neurons in the larval prawn: spontaneous and induced by microlesions. J Neurosci 2:121–142
Fromherz P, Schaden H (1994) Defined neuronal arborizations by guided outgrowth of leech neurons in culture. Eur J Neurosci 6:1500–1504
Fromherz P, Schaden H, Vetter T (1991) Guided outgrowth of leech neurons in culture. Neurosci Lett 129:77–80
Fuchs PA, Nicholls JG, Ready D (1981) Membrane properties and selective connexions of identified leech neurones in culture. J Physiol (Lond) 339:309–324
Fuchs PA, Henderson LP, Nicholls JG (1982) Chemical transmission between individual Retzius and sensory neurones of the leech in culture. J Physiol 323: 195–210
Fuentes EJ, Pachter R, Tsonis PA (1994) On the three-dimensional structure of the hydra head activator neuropeptide In Vivo 8:199–206

Fujisawa T (1989) Role of interstitial cell migration in generating position-dependent patterns of nerve cell differentiation in *Hydra.* Dev Biol 133:77–82
Funte LR, Haydon PG (1993) Presynaptic target contact enhances presynaptic calcium influx by activating cAMP-dependent protein kinase during synaptogenesis. Neuron 10:1069–1078
Gabel CA, Den H, Ambron RT (1989) Characterization of protein-linked glycoconjugates produced by identified neurons of *Aplysia californica.* J Neurobiol 20:530–548
Gallant PE (1988) Effects of the external ions and metabolic poisoning on the constriction of the squid giant axon after axotomy. J Neurosci 8:1479–1484
Gallant PE (1992) The direct effects of graded axonal compression on axoplasm and fast axoplasmic transport. J Neuropathol Exp Neurol 51: 220–230
Gao W-Q, Macagno ER (1987a) Extension and retraction of axonal projections by some developing neurons in the leech depends upon the existence of neighboring homologues. I. The HA cells. J Neurobiol 18:43–59
Gao W-Q, Macagno ER (1987b) Extension and retraction of axonal projections by some developing neurons in the leech depends upon the existence of neighboring homologues. II. The AP and AE neurons. J Neurobiol 18:295–313
Gao W-Q, Macagno ER (1988) Axon extension and retraction by leech neurons: severing early projections to peripheral targets prevents normal retraction of other projections. Neuron 1: 269–277
Garcia-Arraras JE, Torres I, Arroyo E, Cruz W, Jimenez L, Estrada L, Rivera K, Diaz-Miranda L (1994) Neuronal regeneration in *Holothuria glaberrima.* Soc Neurosci Abstr 20:1268
Garcia-Fernàndez J, Bagunà J, Salo E (1991) Planarian homeobox genes: cloning, sequence analysis and expression. Proc Natl Acad Sci USA 88: 7338–7342
Geraerts WPM, Joosse J (1989) Central and peripheral expression of neuropeptide genes: functional and evolutionary perspectives. J Endocrinol Invert 12 (Suppl 2):9
Gertler FB, Bennett RL, Clark MJ, Hoffmann FM (1989) Drosophila *abl* tyrosine kinase in embryonic CNS axons: a role of axonogenesis is revealed through dosage-sensitive interactions with *disabled.* Cell 58:103–113
Ghirardelli E (1965) Differentiation of the germ cells and regeneration of the gonads in planarians. In: Kiortsis V, Trampusch HAL (eds) Regeneration in animals and related problems. North Holland, Amsterdam, pp 177–184
Gillette R (1991) On the significance of neural gigantism in gastropods. Biol Bull 180:234–240
Gilly WF, Brismar T (1989) Properties of appropriately and inappropriately expressed sodium channels in squid giant axon and its somata. J Neurosci 9(4):1362–1374
Glanzman DL, Kandel ER, Schacher S (1989) Identified target motor neuron regulates neurite outgrowth and synapse formation of *Aplysia* sensory neurons in vitro. Neuron 3: 441–445
Glanzman DL, Kandel ER, Schacher S (1990) Target-dependent structural changes accompanying long-term synaptic facilitation in *Aplysia* neurons. Science 249:799–802
Glanzman DL, Kandel ER, Schacher S (1991) Target-dependent morphological segregation of *Aplysia* sensory outgrowth in vitro. Neuron 7:903–913
Glanzman DL, Mackey SL, Hawkins RD, Dyke AM, Lloyd PE, Kandel ER (1989) Depletion of serotonin in the nervous system of *Aplysia* reduces the behavioral enhancement of gill withdrawal as well as the heterosynaptic facilitation produced by tail shock. J Neurosci 9:4200–4213
Glover JC, Mason L (1986) Morphogenesis of an identified leech neuron: segmental specification of axonal outgrowth. Dev Biol 115:256–260
Gnegy ME (1993) Calmodulin in neurotransmitter and hormone action. Annu Rev Pharmacol Toxicol 33:45–70
Go MJ, Hotta Y (1992) An antigen present in the *Drosophila* central nervous system only during embryonic and metamorphic stages. J Neurobiol 23:890–904
Goedert M, Crowther RA, Garner CC (1991) Molecular characterization of microtubule-associated proteins tau and MAP2. Trends Neurosci 14: 193–199
Goldberg DJ, Burmeister DW (1989) Looking into growth cones. Trends Neurosci 12:503–506
Goldberg DJ and Burmeister DW (1992) Microtubule-based filopodium-like protrusions form after axotomy. J Neurosci 12: 4800–4807

Goldberg DJ, Schacher S (1987) Differential growth of the branches of a regenerating bifurcate axon is associated with differential axonal transport of organelles. Dev Biol 124:35–40
Goldberg DJ, Burmeister DW, Rivas RJ (1991) Video microscopic analysis of events in the growth cone underlying axon growth and the regulation of these events by substrate-bound proteins. In: Letourneau PC, Kater SB, Macagno ER (eds) The nerve growth cone. Raven Press, New York, pp 79–95
Goldberg JI, Kater SB (1989) Expression and function of the transmitter serotonin during development of the *Helisoma* nervous system. Dev Biol 131: 483–495
Goldberg JI, McCobb DP, Guthrie PB, Lawton RA, Lee RE, Kater SB (1988) Characterization of cultured embryonic neurons from the snail *Helisoma*. In: Beadle DJ, Lees G, Kater SB (eds) Cell culture approaches to invertebrate neuroscience. Academic Press, Harcourt Brace Javanovich, London, pp 85–108
Goldberg JI, Mills LR, Kater SB (1990) Novel effects of serotonin on neurite outgrowth in neurons cultured from embryos of *Helisoma trivolvis*. J Neurobiol 22: 182–194
Goldberg JI, Mills LR, Kater SB (1991) Novel effects of serotonin on neurite outgrowth in neurons cultured from embryos of *Helisoma trivolvis*. J Neurobiol 22:182–194
Goldberg JI, Mills LR, Kater SB (1992) Effects of serotonin on intracellular calcium in embryonic and adult *Helisoma* neurons. Int J Dev Neurosci 10:255–264
Golding DW (1967a) Regeneration and growth control in *Nereis* I. Growth and regeneration J Embryol Exp Morphol 18:67–77
Golding DW (1967b) Regeneration and growth control in Nereis II. An axial gradient in growth potentiality J Embryol Exp Morphol 18:79–90
Golding DW (1967c) Neurosecretion and regeneration in Nereis I Regeneration and the role of the supraesophageal ganglion. Gen Comp Endocrinol 8:348–355
Golding DW (1967d) Neurosecretion and regeneration in Nereis II. The prolonged secretory activity of the supraesophageal ganglion. Gen Comp Endocrinol 8:356–367
Golding DW (1974a) A survey of neuroendocrine phenomena in non-arthropod invertebrates. Biol Rev 49: 161–224
Golding DW (1974b) Regeneration and growth control in *Nereis* III. Separation of wound healing and segment regeneration by experimental endocrine manipulation. J Embryol Exp Morphol 32:99–109
Golding DW (1987) Brain-body interactions in *Nereis* deactivation of the cerebral neuroendocrine system by ganglion transplantation. Int J Invertebr Reprod Dev 12:281–294
Gomot A, Gomot L, Claude-Roland M, Colard C, Bride J (1992) Immunocytochemical localization of insulin-related peptide(s) in the central nervous system of the snail *Helix aspersa* Muller: Involvement in growth control. Cell Mol Neurobiol 12:21–32
Gonzalez M de L, Silver J (1994) Axon-glia interactions regulate ECM patterning in the postnatal rat olfactory bulb. J Neurosci 14: 6121–6131
Goodman CS (1994) The likeness of being: phylogenetically conserved molecular mechanisms of growth cone guidance. Cell 78: 353–356
Goodman CS, Heitler WJ (1979) Electrical properties of insect neurones with spiking and non-spiking somata: normal, axotomized and colchicine-treated neurones. J Exp Biol 83:95–121
Goodman CS, Shatz CJ (1993) Developmental mechanisms that generate precise patterns of neuronal connectivity. Neuron (Suppl) 10:77–98
Goransson LG, Hunt WP, Vélez SJ (1988) Regeneration studies on a crayfish neuromuscular system. II. Effect of changing the nerve entry point into the muscle field on the gradient of innervation. J Neurobiol 19:141–152
Gordon-Weeks PR (1991) Evidence for microtubule capture by filopodial actin filaments in growth cones. NeuroReport 2: 573–576
Gordon-Weeks PR, Mansfield SG (1990) Assembly of microtubules in growth cones: the role of microtubule-associated proteins. In: Letourneau PC, Kater SB, Macagno ER (eds), The nerve growth cone. Raven Press, New York, pp 55–64
Goslin F, Schreyer DJ, Skene JHP, Banker G (1988) Development of neuronal polarity: GAP-43 distinguishes axonal from dendritic growth cones. Nature 336:672–674
Goss RJ (1969) Principles of regeneration. Academic Press, New York

Gottmann K, Rohrer H, Lux HD (1991) Distribution of Ca^{2+} and Na+ conductances during neuronal differentiation of chick DRG precursor cells. J Neurosci 11: 3371–3378
Govind CK (1992) Nervous system. Micro Anat Invert 10:395–438
Govind CK, Pearce J, Potter DJ (1988) Neural attrition following limb loss and regeneration in juvenile lobsters. J Neurobiol 19: 667–680
Govind CK, Blundon JA, Kirk MD (1992) Functional degeneration of isolated central stumps of crayfish sensory axons. J Comp Neurol 322:111–120
Graf RA, Cooke IM (1990) Primary culture of crustacean stomatogastric ganglion neurones in a defined medium. J Exp Biol 149:521–525
Graf RA, Cooke IM (1994) Outgrowth morphology and intracellular calcium of crustacean neurons displaying distinct morphologies in primary culture. J Neurobiol 25: 1558–1569
Gras H (1984) The tail flattening reflex of *Lumbricus*: reconstitution after tail amputation and modifications in segmental nerve roots. J Neurobiol 15: 249–261
Grau SM, Cooke IM (1992) Peptidergic neurons of the crab, *Cardisoma carnifex*, in defined culture maintain characteriatic morphologies under a variety of conditions. Cell Tissue Res 270: 303–317
Graziadei PPC, Monti-Graziadei GA (1978) Continuous nerve cell renewal in the olfactory system. In: Jacobsonm M (ed) Handbook of sensory physiology, vol IX. Springer, Berlin Heidelberg, New York, pp 55–83
Graziadei PPC, Monti-Graziadei GA (1980) Neurogenesis and neuron regeneration in the olfactory system of mammals. III Differentiation and reinnervation of the olfactory bulb following section of the fila olfactoria in rat. J Neurocytol 9: 145–162
Greger V, Berking S (1991) Nerve cell production in *Hydra* is deregulated by tumour-promoting phorbol ester. Roux's Arch Dev Biol 200:234–236
Grenningloh G, Beiber AJ, Rhem EJ, Snow PM, Traquina ZR, Hortsch M, Patel NH, Goodman CS (1990) Molecular genetics of neuronal recognition in *Drosophila*: evolution and function of immunoglobulin superfamily cell adhesion molecules. Cold Spring Harbor Symp Quant Biol 55: 327–340
Grigoriev N, Spencer AN (1994) Conduction in a pure epithelial sheet: velocity decrease with repetitive stimulation, a case of non-neuronal habituation? Soc Neurosci Abstr 20:582
Grimm-Jorgensen Y (1983) Immunoreactive somatostatin in two pulmonate gastropods. Gen Comp Endocrinol 49:108–114
Grimm-Jorgensen Y (1987) Somatostatin and calcitonin stimulate regeneration of molluscan neurons in vitro. Brain Res 403:121–126
Grinvald A, Farber I (1981) Optical recording of calcium action potentials from growth cones of cultured neurons using a laser microbeam. Science 212: 1164–1169
Grossfeld RM, Hansen GD (1987) Long-term persistence of GAD activity in injured crayfish CNS tissue. Neurochem Res 12:977–983
Grossfeld RM, McKinnon E, Hargittai PT, Lieberman EM (1994) Glutamine cycle enzymes in the crayfish medial giant nerve fiber: implications for axon-to-glia signaling. Soc Neurosci Abs 20:914
Grumbacher-Reinert S (1989) Local influence of substrate molecules in determining distinctive growth patterns of identified neurons in culture. Proc Natl Acad Sci USA 86:7270–7274
Grumbacher-Reinert S, Nicholls J (1992) Influence of substrate on retraction of neurites following electrical activity of leech retzius cells in culture. J Exp Biol 167:1–14
Gu X, Muller KJ (1990) Competitive interactions between neurons making axosomatic contacts in the leech. J Neurosci 10:3814–1822
Gustafson T, Toneby M (1970) On the role of serotonin and acetylcholine in sea urchin morphogenesis. Exp Cell Res 62:102–117
Gustafsson MKS (1992) The neuroanatomy of parasitic flatworms. Adv Neuroimmun 2: 267–286
Gustafsson MKS, Reuter M (1990) Flatworms as model organisms. In: Gustafsson MKS, Reuter M (eds) The early brain. Proc Symp Invertebrate neurobiology. Acta Acad Abo Ser B 50:5–12
Gustafsson MKS, Eriksson K (1992) Never ending growth and a growth factor. I Immunocytochemical evidence for the presence of basic fibroblast growth factor in a tapeworm. Growth Factors 7:327–334

Guthrie DM (1967) The regeneration of motor axons in an insect. J Insect Physiol 13:1593–1611
Guthrie SC (1985) Cell migration in the patterning of the nervous system of *Hydra*. Trends Neurosci 8: 306–308
Gutmann E, Guttmann L, Medawar PB, Young JZ (1941) The rate of regeneration of nerve. J Exp Biol 18:14–44
Hadley RD, Wong RG, Kater SB, Barker DL, Bulloch AGM (1982) Formation of novel central and peripheral connections between molluscan central neurons in organ cultured ganglia. J Neurobiol 13: 217–230
Hadley RD, Kater SB, Cohan CS (1983) Electrical synapse formation depends on interaction of mutually growing neurites. Science 221:466–468
Hall DH, Russell RL (1991) The posterior nervous system of the nematode *Caenorhabditis elegans*: serial reconstruction of identified neurons and complete pattern of synaptic interactions. J Neurosci 11:1–22
Hall ZW, Sanes JR (1993) Synaptic structure and development: the neuromuscular junction. Cell 72 (Suppl):99–122
Hall JD, Lloyd PE (1991) Release of pedal peptide from *Aplysia* neurons in primary culture. J Neurobiol 22:858–868
Hall FG, Poulos A, Cohen MJ (1989) Sprouts emerging from the dendrites of axotomized lamprey central neurons have axon-like ultrastructure. J Neurosci 9:588–599
Halpern ME, Chiba A, Johansen J, Keshishian H (1991) Growth cone behavior underlying the development of stereotypic synaptic connections in *Drosophila* embryos. J Neurosci 11:3227–3238
Hamamoto S, Imagawa W, Yang J, Nandi S (1988) Morphogenesis of mouse mammary epithelial cells growing within collagen gels: ultrastructural and immunocytochemical characterization. Cell Differ 22:191–202
Hamelin M, Zhou Y, Su MW, Scott, IM, Culotti JG (1993) Expression of the UNC-5 guidance receptor in the touch neurons of *C. elegans* steers their axons dorsally. Nature 364:327–330
Haralson JV, Groff CI (1975) Classical conditioning in the sea anemone, *Cribina xanthogrammica*. Physiol Behav 15:455–460
Harrelson AL, Goodman CS (1988) Growth cone guidance in insects: fasciclin II is a member of the immunoglobulin superfamily. Science 242: 700–708
Harzsch S, Dawirs RR (1994) Neurogenesis in larval stages of the spider crab *Hyas araneus* (Decapoda, Brachyura): proliferation of neuroblasts in the ventral nerve cord. Roux's Arch Dev Biol 204:93–100
Hauser M, Koopowitz H (1987) Age-dependent changes in fluorescent neurons in the brain of *Notoplana acticola*, a polyclad flatworm. J Exp Zool 241:217–225
Hawver DB, Schacher S (1993) Selective fasciculation as a mechanism for the formation of specific chemical connections between *Aplysia* neurons in vitro. J Neurobiol 24: 368–383
Hay ED (1966) Regeneration. Holt, Rinehart and Winston, New York
Haydon PG (1988) The formation of chemical synapses between cell-cultured neuronal somata. J Neurosci 8:1032–1038
Haydon PG, Kater SB (1988) The differential regulation of formation of chemical and electrical connections in *Helisoma*. J Neurobiol 19: 636–655
Haydon PG, Zoran MJ (1989) Formation and modulation of chemical connections: evoked acetylcholine release from growth cones and neurites of specific identified neurons. Neuron 2:1483–1490
Haydon PG, Zoran MJ (1994) Retrograde regulation of presynaptic development during synaptogenesis. J Neurobiol 25: 694–706
Haydon PG, McCobb DP, Kater SB (1984) Serotonin selectively inhibits growth cone motility and synaptogenesis of specific identified neurons. Science 226: 561–564
Haydon PG, Cohan CS, McCobb DP, Miller HR, Kater SB (1985) Neuron-specific growth cone properties as seen in identified neurons of *Helisoma*. J Neurosci Res 13: 135–147
Haydon PG, McCobb DP, Kater SB (1987) The regulation of neurite outgrowth growth cone motility and electrical synaptogenesis by serotonin. J Neurobiol 18:197–215
Hedgecock EM, Culotti JG, Hall DH, Stern BD (1987) Genetics of cell and axon migrations in *Caenorhabditis elegans*. Development 100:365–382

Hedgecock EM, Culotti JG, Hall DH (1990) The *unc-5, unc-6, and unc-40* genes guide circumferential migrations of pioneer axons and mesodermal cells on the epidermis in *C. elegans.* Neuron 2: 61–85
Hekimi S, Kershaw D (1993) Axonal guidance defects in a *Caenorhabditis elegans* mutant reveal cell-extrinsic determinants of neuronal morphology. J Neurosci 13(10):4254–4271
Henderson LP (1983) The role of 5-hydroxytryptamine as a transmitter getween identified leech neurones in culture. J Physiol 339: 309–324
Herlant-Meewis H (1962) Neurosecretory phenomena during regeneration of nervous centres in *Eisenia foetida.* Mem Soc Endocr 12:267–274
Herlant-Meewis H (1964) Regeneration in annelids. In: Abercrombe M, Brachet J (eds) Advances in morphogenesis, vol 4. Academic Press, New York, pp 155–215
Hernandez-Nicaise M-L (1973) Le système nerveux des Cténaires. I. Structure et ultrastrudture des réseaux epithéliaux. Z Zellforsch 137:223–250
Herrup K, Busser J (1994) Target-related cell death is associated with reentry into the cell cycle. Soc Neurosci Abstr 20:245
Hickmott PW, Carew TJ (1991) An autoradiographic analysis of neurogenesis in juvenile *Aplysia californica.* J Neurobiol 22:313–326
Hildebrand JG (1985) Metamorphosis of the insect nervous system. Influences of the periphery on the postembryonic development of the antennal sensory pathway in the brain of *Manduca sexta.* In: Selverston A (ed) Model neural networks and behavior. Plenum, New York pp 279–285
Hinrichsen RD (1993) Calcium and calmodulin in the control of cellular behavior andmotility. Biochem Biophys Acta 1155:277–293
Hitchcock PF, Raymond PA (1992) Retinal regeneration. Trends Neurosci 15:103–108
Holder N, Clark JDW (1988) Is there a correlation between continuous neurogenesis and directed axon regeneration in the vertebrate nervous system? Trends Neurosci 11:94–99
Holliday J, Spitzer NC (1990) Spontaneous calcium influx and its roles in differentiation of spinal neurons in culture. Dev Biol 141:13–23
Holstein TW, David CN (1990) Putative intermediates in the nerve cell differentiation pathway in hydra have properties of multipotent stem cells. Dev Biol 142:401–405
Hopkins PM, Bliss DE, Sheehan SW, Boyer JR (1979) Limb growth-controlling factors in the crab *Gecarcinus lateralis,* with special reference to the limb growth-inhibiting factor. Gen Comp Endocrinol 39: 192–207
Hortsch M, Goodman CS (1991) Cell and substrate adhesion molecules in *Drosophila.* Annu Rev Cell Biol 7:505–557
Howes EA, Smith PJS, Treherne JE (1987) Glial repair in the cultured central nervous system of an insect. Cell Tiss Res 247:111–120
Howes EA, Smith PJS, Treherne JE (1989) Adult insect glial culture: activation, substrate effects and proliferation. Tissue Cell 21: 759–772
Howes EA, Cheek TR, Smith PJS (1991) Long-term growth in vitro of isolated, fully differentiated neurones from the central nervous system of an adult insect. J Exp Biol 156:591–605
Hoy RR, Bittner GD, Kennedy D (1967) Regeneration in crustacean neurons: evidence for axonal fusion. Science 15:251–252
Huber F (1987) Plasticity in the auditory system of crickets: phonotaxis with one ear and neuronal reorganization within the auditory pathway. J Comp Physiol 161A:583–604
Huber F, Thorson J (1985) Cricket auditory communication. Sci Am 253:60–68
Huet M (1975) Le role du systeme nerveux au cours de la regeneration du bras chez le etoile de mer: *Asterina gibbosa* Penn. (Echinoderme, asteridae). J Embryol Exp Morphol 33:535–552
Hulsebosch CE, Bittner GD (1980) Evolution of abilities to regenerate neurons in central nervous systems. Am Nat 115:276–284
Hume RI, Role LW, Fishbach GD (1983) Acetylcholine release from growth cones detected with patches of acetylcholine receptor-rich membranes. Nature 305:632–634
Hunt WP, Vélez SJ (1982) Regeneration of specific neuromuscular connections in the crayfish. II. Effect of changes in target area. J Neurophysiol 47:666–676
Hunter DD, Cashman N, Morris-Valero R, Buolck JW, Adams SP, Sanes JR (1991) An LRE (leucine-arginine-glutamate)-dependent mechanism for adhesion of neurons to S-laminin. J Neurosci 11:3960–3971

Ito K, Hotta Y (1992) Proliferation pattern of postembryonic neuroblasts in the brain of *Drosophila melanogaster.* Dev Biol 149: 134–148
Ivins JK, Raper JA, Pittman RN (1991) Intracellular calcium levels do not change during contact-mediated collapse of chick DRG growth cone structure. J Neurosci 11: 1597–1608
Jacklet JW, Cohen MJ (1967) Nerve regeneration: correlation of electrical, histological and behavioral events. Science 156: 1640–1643
Jacob MH (1984) Neurogenesis in *Aplysia californica* resembles nervous system formation in vertebrates. J Neurosci 4:1225–1239
Jacob JM, McQuarrie IG (1991) Axotomy accelerates the slow component b of axonal transport. J Neurobiol 22:570–582
Jacob JM, McQuarrie IG (1993) Acceleration of axonal outgrowth in rat sciatic nerve at one week after axotomy. J Neurobiol 24: 356–367
Jansen JKS, Nicholls JG (1972) Regeneration and changes in synaptic connections between individual nerve cells in the central nervous system of the leech. Proc Natl Acad Sci USA 69:636–639
Jansen JKS, Muller KJ, Nicholls JG (1974) Persistent modification of synaptic interactions between sensory and motor nerve cells following discrete lesions in the central nervous system of the leech. J Physiol (Lond) 242:289–305
Jay DG, Keshishian H (1990) Laser inactivation of fasciclin I disrupts axon adhesion of grasshopper pioneer neurons. Nature 348: 548–550
Jellies J, Loer CM, Kristan WB, Jr (1987) Morphological changes in leech Retzius neurons after target contact during leech embryogenesis. J Neurosci 7:2618–2629
Jellies J, Johansen K, Johansen J (1994) Specific pathway selection by the early projections of individual peripheral sensory neurons in the embryonic medicinal leech. J Neurobiol 25:1187–1199
Jeppesen LL (1976) The control of mating behavior in *Helix pomatia* L. (Gastropoda: Pulmonata). Anim Behav 24: 275–290
Jessell TM (1988) Adhesion molecules and the hierarchy of neural development. Neuron 1:3–13
Jessell TM, Hynes MA, Dodd J (1990) Carbohydrates and carbohydrate-binding proteins in the nervous system. Annu Rev Neurosci 13:227–255
Johansen J, Thompson I, Stewart RR, McKay RG (1985) Expression of surface antigens recognized by the monoclonal antibody Lan 3–2 during embryonic development in the leech. Brain Res 343: 1–7
Johansen KM, Kopp DM, Jellies J, Johansen J (1992) Tract formation and axon fasciculation of molecularly distinct peripheral neuron subpopulations during leech embryogenesis. Neuron 8:559–572
Jones PG, Bulloch AGM (1988) L-glutamate promotes neurite outgrowth in adult *Helisoma* neurons. Neurosci Res Commun 3:93–98
Jones PG, Rosser SJ, Bulloch AGM (1987) Glutamate suppression of feeding and the underlying output of effector neurons in *Helisoma.* Brain Res 437:56–68
Kaczmerek LK, Finbow M, Revel JP, Strumvasser F (1979) The morphology and coupling of *Aplysia* bag cells within the abdominal ganglion and in cell culture. J Neurobiol 10:535–550
Kalmring K (1975) The afferent auditory pathway in the ventral cord of *Locusta migratoria* (Acrididae). I. Synaptic connectivity and information processing among the auditory neurons of the ventral cord. J Comp Physiol 104:103–141
Kämper G, Murphey RK (1987) Synapse formation by sensory neurons after cross-species transplantation in crickets: the role of positional information. Dev Biol 122:492–502
Karlstrom RO, Wilder LP, Bastiani MJ (1993) Lachesin: an immunoglobulin superfamily protein whose expression correlates with neurogenesis in grasshopper embryos. Development 118:509–522
Kater SB (1985) Dynamic regulators of neuronal form and connectivity in the adult snail *Helisoma.* In: Selverston AI (ed) Model neural networks and behavior, Plenum, New York, pp 191–209
Kater SB, Mattson MP (1988) Extrinsic and intrinsic regulators of neurite outgrowth and synaptogenesis in isolated, identified *Helisoma* neurons in culture. In: Beadle DJ, Lees G, Kater SB (eds) Cell culture approaches to invertebrate neuroscience, Acad Press, London, pp 1–31

Kater SB, Mills LR (1991) Regulation of growth cone behavior by calcium. J Neurosci 11:891–899
Keller F, Schacher S (1990) Neuron-specific glycoproteins promoting neurite fasciculation in *Aplysia californica*. J Cell Biol 111: 2637–2650
Kempf SC, Chun GV, Hadfield MG (1992) An immunocytochemical search for potential neurotransmitters in larvae of *Phestilla sibogae* (Gastropoda, Opisthobranchia). Comp Biochem Physiol 101C:299–305
Kennedy D (1971) Crayfish interneurons. Physiologist 14:5–30
Kennedy D, Bittner GD (1974) Ultrastructural correlates of motor nerve regeneration in crayfish. Cell Tissue Res 148:97–110
Keshishian H (1980) The origin and morphogenesis of pioneer neurons in the grasshopper metathoracic leg. Dev Biol 80:388–397
Keshishian H (1993) Making the right connections. Nature 361:299–300
Keshishian H, Chang TN, Jarecki J (1994) Precision and plasticity during *Drosophila* neuromuscular development. FASEB J 8:731–737
Keynes R, Cook G (1990) Cell-cell repulsion cues from the growth cone? Cell 62:609–610
Killian KA, Merritt DJ, Murphey RK (1993) Transplantation of neruons reveals processing areas and rules for synaptic connectivity in the cricket nervous system. J Neurobiol 24: 1187–1206
Kim GJ, Shatz CJ, McConnell SK (1991) Morphology of pioneer and follower growth cones in the developing cerebral cortex. J Neurobiol 22: 629–642
Kim YT, Wu C-F (1991) Distinctions in growth cone morphology and motility between monopolar and multipolar neurons in *Drosophila* CNS cultures. J Neurobiol 22:263–275
Kimble J (1981) Alterations in cell lineage following laser ablation of cells in the somatic gonad of *Caenorhabditis elegans*. Dev Biol 87:286–300
Kimmel HD, Carlyon WD (1990) Persistent effects of a serotonin depletor (*p*-chlorophenylalanine) in regenerated planaria (*Dugesia dorotocephala*). Behav Neurosci 104:127–134
Kiortsis V, Moraitou M (1965) Factors of regeneration in *Spirographis spallanzanii*. In: Kiortsis V, Trampusch HAL (eds) Regeneration in animals and related problems. North Holland, New York, pp 250–261
Kiortsis V, Trampusch HAL (1965) Regeneration in animals and related problems. North Holland, New York
Kirk MD, Govind CK (1992) Early innervation of abdominal swimmeret muscles in developing lobsters. J Exp Zool 261:298–309
Kits KS, de Vries NJ, Ebberink RHM (1990) Molluscan insulin-related neuropeptide promotes neurite outgrowth in dissociated neuronal cell cultures. Neurosci Lett 109:253–258
Kleinfeld D, Kahler KH, Hockberger PE (1988) Controlled outgrowth of dissociated neurons on patterned substrates. J Neurosci 8:4098–4120
Kleinfeld D, Raccuia-Behling F, Chief HJ (1990) Circuits constructed from identified *Aplysia* neurons exhibit multiple patterns of persistent activity. Biophys J 57:697–715
Klose M, Bentley D (1989) Transient pioneer neurons are essential for formation of an embryonic peripheral nerve. Science 245:982–983
Knobloch CA, Steel CGH (1988) Interactions between limb regeneration and ecdysteriod titres in last larval instar *Rhodnius prolixus* (Hemiptera). J Insect Physiol 34: 507–514
Koizumi O, Bode HR (1991) Plasticity in the nervous system of adult hydra. III Conversion of neurons to expression of a vasopressin-like immunoreactivity depends on axial location. J Neurosci 11:2011–2020
Koizumi O, Wilson JD, Grimmelikhuijzen CJP, Westfall JA (1989) Ultrastructural localization of RFamide-like peptides in neuronal dense-cored vesicles in the peduncle of *Hydra*. J Exp Zool 249: 17–22
Koizumi O, Mizumoto H, Sugiyama T, Bode HR (1990) Nerve net formation in the primitive nervous system of *Hydra* – an overview. Neurosci Res (Suppl) 13:S165-S170
Kolodkin AL, Matthes D, O'Connor TP, Patel NH, Admon A, Bentley D, Goodman CS (1992) Fasciclin IV: sequence, expression and function during growth cone guidance in the grasshopper embryo. Neuron 9:831–845
Koopowitz H (1989) Polyclad neurobiology and the evolution of central nervous systems. In: Anderson PAV (ed) Evolution of the first nervous systems. Plenum Press, New York, pp 3151–328

Koopowitz H, Holman M (1988) Neuronal repair and recovery of function in the polyclad flatworm, *Notoplana acticola*. Am Zool 28:1065–1075
Koopowitz H, Silver D, Rose G (1975) Neuronal plasticity and recovery of function in a polyclad flatworm. Nature 256:737–738
Koopowitz H, Silver D, Rose G (1976) Primitive nervous systems. Control and recovery of feeding behavior in the polyclad flatworm, *Notoplana acticola*. Biol Bull 150:411–425
Koyama H, Kusunoki T (1993) Organization of the cerebral ganglion of the colonial ascidian *Polyandrocarpa misakiensis*. J Comp Neurol 338:549–559
Krasne FB (1987) Silencing normal input permits regenerating foreign afferents to innervate an identified crayfish sensory interneuron. J Neurobiol 18:61–73
Krause TL, Bittner GD (1990) Rapid morphological fusion of severed myelinated axons by polyethylene glycol. Proc Natl Acad Sci USA 87:1471–1475
Krause TL, Fishman HM, Ballinger ML, Bittner GD (1994) Extent and mechanism of sealing in transected giant axons of squid and earthworms. J Neurosci 14:6638–6651
Krause TL, Marquis RE, Lyckman AW, Ballinger ML, Bittner GD (1991) Rapid artificial restoration of electrical connections across a crush lesion of a giant axon. Brain Res 561:350–353
Krishnan SN, Frei E, Swain GP, Wyman RJ (1993) *Passover*: a gene required for synaptic connectivity in the giant fiber system of *Drosophila*. Cell 73:967–977
Kristan WB, French KA, Szczupak L (1993) Developmental regulation of segment-specific cholinergic receptors on retzius neurons in the medicinal leech. J Neurosci 13(4):1577–1587
Kruk PJ, Bulloch AGM (1992) Axonal regeneration of an identified *Helisoma* neuron depends on the site of axotomy. J Neurosci Res 31:401–412
Krull CE, Morton DB, Faissner A, Schachner M, Tolbert LP (1994) Spatiotemporal pattern of expression of tenascin-like molecules in a developing insect olfactory system. J Neurobiol 25:515–534
Kuno M, Llinas R (1970) Enhancement of synaptic transmission by dendritic potentials in chromatolysed motoneurones of the cat. J Physiol 210:807–821
Kupfermann I (1974) Dissociation of the appetive and consummatory phases of feeding behavior in *Aplysia*: a lesion study. Behav Biol 10:89–97
Kuwada JY (1981) Ionic and metabolic dependence of axotomy-induced somatic membrane changes. J Physiol (Lond) 317:463–473
Kuwada JY, Wine JJ (1981) Transient, axotomy-induced changes in the membrane properties of crayfish central neurones. J Physiol Lond 317:435–461
LaBerge S, Chase R (1992) The development of mesocerebral neurons in the snail *Helix aspersa maxima*. Can J Zool 70:2034–2041
Lakes R (1990) Plasticity in the nervous system of orthopterans. In: Gribakin FG, Wiese K, Popov AV (eds) Sensory systems and communication in arthropods. Birkhauser Verlag, Basel, pp 280–284
Lakes R, Kalmring K, Engelhard K-H (1990) Changes in the auditory system of locusts (*Locusta migratoria* and *Schistocerca gregaria*) after deafferentation. J Comp Physiol 166A:553–563
Lakes-Harlan R, Pollack GS (1993) Pathfinding of peripheral neurons in the central nervous system of an embryonic grasshopper (*Chorthippus biguttulus*). Cell Tissue Res 273:97–106
Lander AD (1989) Understanding the molecules of neural cell contacts: emerging patterns of structure and function. Trends Neurosci 12:189–195
Landmesser L, Dahm L, Tang J, Rutishauser U (1990) Polysialic acid as a regulator of intramuscular nerve branching during embryonic development. Neuron 4:655–667
Lange MM (1920) On the regeneration and finer structure of the arms of the cephalopods. J Exp Zool 31:1–57
Lankford KL Letourneau P (1989) Evidence that calcium may control neurite outgrowth by regulating the stability of actin filaments. J Cell Biol 109: 1229–1243
Lasek RJ, Gainer H, Przybylski RJ (1974) Transfer of newly synthesized proteins from Schwann cells to the squid giant axon. Proc Natl Acad Sci USA 71: 1188–1192
Lauder JM (1993) Neurotransmitters as growth regulatory signals: role of receptors and second messengers. Trends Neurosci 16:233–239

Leah JD, Herdegen T, Bravo R (1991) Selective expression of Jun proteins following axotomy and axonal transport block in peripheral nerves in the rat: evidence for a role in the regeneration process. Brain Res 566:198–207
Lees G, Beadle DJ (1991) Cockroach neuronal cultures as models for investigating the neuropharmacology of the insect central nervous system. In: Beadle DJ, Lees G, Kater SB (eds) Cell culture approaches to invertebrate neuroscience. Academic Press, New York, pp 109–147
Le Gall P (1983) The neuroendocrine control of cell multiplication in the regenerating tentacles and the germinal cells of *Crepidula fornicata*. In: Lever J, Boer HH (eds) Molluscan neuroendocrinology. North Holland, Amsterdam, pp 175–177
Le Gall S, Le Gall P, Feral C (1983) The neuroendocrine mechanism responsible for penis differentiation in *Crepidula fornicata* L. (protandric hermaphrodite Mesogastropod). In: Lever J, Boer HH (eds) Molluscan neuroendocrinology. North Holland, Amsterdam pp 169–173
Lender T (1956) Recherches experimentales sur la nature et les proprietes de l'inducteur de la regeneration des yeux de la planaire *Polycelis nigra*. J Embryol Exp Morphol 4: 196–216
Lender T, Bouchard-Madrell C (1964) Etudes experimentale de la regeneration du complex neural de *Ciona intestinalis* (Prochorde). Bull Soc Zool 89: 546–554
Lender T, Bouchard-Madrelle C (1964) Etudes experimentale de la regeneration du complex neural de *Ciona intestinalis* (Prochorde). Bull Soc Zool 89:546–554
Letourneau JG (1976) Addition of sensory structures and associated neurons to the crayfish telson during development. J Comp Physiol 110: 13–23
Letourneau PC, Kater SB, Macagno ER (1991) The growth cone. Raven Press, New York
Lichtman JW (1982) Factors involved in the restoration of specific neural connections – group report. In: Nicholls JG (ed) Repair and regeneration of the nervous system. Dahlem Konferenzen. Springer, Berlin, pp 343–379
Lichtman JW, Balice-Gordon RJ (1989) Understanding synaptic competition in theory and practice. J Neurobiol 21:99–106
Lieberman AR (1971) The axon reaction: a review of the principal features of of perikaryal response to axon injury. Int Rev Neurobiol 12:49–124
Lin SS, Levitan IB (1987) Concanavalin A alters synaptic specificity between cultured *Aplysia* neurons. Science 237: 648–650
Lin SS, Sammak PJ, Borisy GG (1989) Progressive and spatially differentiated stability of microtubules in developing neuronal cells. J Cell Biol 111: 495–509
Lin SS, Dagan D, Levitan IB (1989) Concanavalin A modulates a potassium channel in cultured *Aplysia* neurons. Neuron 3: 95–102
Lnenicka GA, Murphey RK (1989) The refinement of invertebrate synapses during development. J Neurobiol 20:339–355
Lnenicka GH, Atwood HL (1989) Long-term changes in neuromuscular synapses with altered sensory input to a crayfish motor neuron. Exp Neurol 100:437–447
Lnenicka GA, Hong SJ, Combatti M, LePage S (1991) Activity-dependent development of synaptic varicosities at crayfish motor terminals. J Neurosci 11:1040–1048
Loer CM, Kristan WB Jr (1989) Central synaptic inputs to identified leech neurons determined by peripheral targets. Science 244:64–66
Loer CM, Jellies J, Kristan WB Jr (1987) Segment-specific morphogenesis of leech Retzius neurons requires particular peripheral targets. J Neurosci 7:2630–2638
Lohof AM, Quillan M, Dan Y, Poo M (1992) Asymmetric modulation of cytosolic cAMP activity induces growth cone turning. J Neurosci 12: 1253–1261
LoPresti V, Macagno ER, Levinthal C (1973) Structure and development of neuronal connection in isogenic organisms: cellular interactions in the development of the optic lamina of *Daphnia*. Proc Natl Acad Sci USA 70: 433–437
LoPresti V, Macagno ER, Levinthal C (1974) Structure and development of neuronal connections in isogenic organisms: transient gap junctions between growing optic axons and lamina neuroblasts. Proc Natl Acad Sci USA 71:1098–1102
Lubet P (1971) Influence des ganglions cerebroides sur la croissance de *Crepidula fornicata* Phil. (Mollusque mesogasteropode). C R Acad Sci Paris 273:2309–2311

Lyckman AW, Bittner GD (1992) Axonal conduction and electrical coupling in regenerating earthworm giant axons. Exp Neurol 117:299–306

Lyckman AW, Heidelbaugh SM, Bittner GD (1992) Analysis of neuritic outgrowth from severed giant axons in *Lumbricus terrestris*. J Comp Neurol 318:426–438

Lyngdoh RD, Tandon V (1992) Organization of nervous system in *Lytocestus indicus* (Cestoda: Caryophyllidea). Zool Anz 228 (5/6):238–247

Macagno ER, Muller KJ, DeRiemer SA (1985) Regeneration of axons and synaptic connections by touch sensory neurons in the leech central nervous system. J Neurosci 5:2510–2521

Macagno ER, Peinado A, Stewart RR (1986) Segmental differentiation in the leech nervous system; specific phenotypic changes associated with ectopic targets. Proc Nat Adac Sci USA 83:2746–2750

Macagno ER, Gao W-Q, Baptista CA, Passani MB (1989) Competition or inhibition? Developmental strategies in the establishment of peripheral projections by leech neurons. J Neurobiol 21:107–119

Mackie GO (1980) Slow swimming and cyclic 'fishing' behavior in *Aglantha digitale* (Hydromedusae Trachylina). Can J Fish Aquat Sci 37:1550–1556

Maetzold DJ, Bulloch AGM (1986) Sprouting by undamaged adult molluscan neurones: putative role for changes in haemolymph osmoregulation. J Exp Biol 122:427–432

Mahanthappa NK (1994) Regeneration in the auditory system: lessons from other epithelia, and persisting puzzles. Trends Neurosci 17:357–359

Marcum BA, Campbell RD (1978) Development of *Hydra* lacking nerve and interstitial cells. J Cell Sci 29:17–33

Marom S, Dagan D (1987) Calcium current in growth balls from isolated *Helix aspersa* neuronal growth cones. Pfluegers Arch Eur J Physiol 409:578–581

Martelly I, Franquinet R (1984) Planarian regeneration as a model for cellular activation studies. Trends Biochem Sci 8: 468–471

Martindale MQ, Shankland M (1990) Neuronal competition determines the spatial pattern of neuropeptide expression by identified neurons of the leech. Dev Biol 139:210–226

Mason A, Muller KJ (1983) Regeneration and plasticity of neuronal connections in the leech. Trends Neurosci 21:172–176

Masuda-Nakagawa LM, Nicholls JG (1991) Extracellular matrix molecules in development and regeneration of the leech CNS. Philos Trans R Soc Lond B Biol Sci 331:323–335

Masuda-Nakagawa LM, Wiedemann C (1992) The role of matrix molecules in regeneration of leech CNS. J Neurobiol 23:551–567

Masuda-Nakagawa L, Beck K, Chiquet M (1988) Identification of cells of molecules in leech extracellular matrix that promote neurite outgrowth. Proc R Soc Lond B Biol Sci 235:247–257

Masuda-Nakagawa LM, Muller KJ, Nicholls JG (1990) Accumulation of laminin and microglial cells at sites of injury and regeneration in the central nervous system of the leech. Proc R Soc Lond B Biol Sci 241:201–206

Masuda-Nakagawa LM, Woulz A, Brodbeck D, Neely MD, Grumbacher-Reinert S (1994) Substrate-dependent interactions of leech microgial cells and neurons in culture. J Neurobiol 25:83–91

Mattson MP, Kater SB (1987) Calcium regulation of neurite elongation and growth cone motility. J Neurosci 7: 4034–4043

Mattson MP, Kater SB (1988) Fibronectin-like immunoreactivity in *Helisoma* buccal ganglia: evidence that an endogenous fibronectin-like molecule promotes neurite outgrowth. J Neurobiol 19:239–256

Mattson MP, Dou P, Kater SB (1988a) Outgrowth-regulating aspects of glutamate in isolated hippocampal pyramidal neurons. J Neurosci 8:2087–2100

Mattson MP, Guthrie PB, Kater SB (1988b) Components of neurite outgrowth that determine neuronal cytoarchitecture: influence of calcium and the growth substrate. J Neurosci Res 20:331–345

Matzner O, Devor M (1994) Hyperexcitability at sites of nerve injury depends on voltage-sensitive Na^+ channels. J Neurophysiol 72: 349–359

Maule AG, Halton DW, Shaw C, Johnston CF (1993) The cholinergic, serotoninergic and peptidergic components of the nervous system of *Moniezia expansa* (Cestoda, Cyclophyllidea). Parasitica 106:429–440

May RH, Ridgway RL, Moffett SB (1987) Size and number of serotonin immunoreactive cells change with age in the adults of the snail *Melampus*. Soc Neurosci Abstr 134:1070

May RM (1933) La formation des terminaisons nerveuses dans les ventouses du bras regenere du cephalopode *Octopus vulgaris* Lam. Ann Sta Oceanogr Salammbo 7:1–15

Mayford M, Barzilai A, Keller F, Schacher S, Kandel ER (1992) Modulation of an NCAM-related adhesion molecule with long-term synaptic plasticity in *Aplysia*. Science 256:638–344

McAllister LB, Scheller RJ, Kandel ER, Axel R (1983) In situ hybridization to study the origin and fate of identified neurons. Science 222:800–808

McCarthy MI, Rubin LL, Philpott KL (1994) Apoptosis: links with the cell cycle? Soc Neurosci Abstr 20:245

McClellan AD (1992) Functional regeneration and recovery of locomotor activity in spinally transected lamprey. J Exp Zool 261:274–287

McCobb DP, Hayden PG, Kater SB (1988a) Dopamine and serotonin regulation of neurite outgrowth of different identified neurons. J Neurosci Res 19:19–26

McCobb DP, Cohan CS, Conner JA, Kater SB (1988b) Interactive effects of serotonin and acetylcholine on neurite elongation. Neuron 1:377–385

McFaddan NA, Spencer AN (1992) The possible involvement of ARG-Phe-amide immunoreactive neurons during tentacle development and regeneration in the hydromedusa *Polyorchis penicillatus*. Can J Zool 70: 567–577

McFarlane ID, Graff D, Grimmelikhuijzen CJP (1989) Evolution of conducting systems and neurotransmitters in the anthozoa. In: Anderson PAV (ed) Evolution of the first nervous systems. NATO Adv Worksh 188. Plenum Press, New York, pp 111–127

McGlade-McCulloh E, Muller KJ (1989) Developing axons continue to grow at their tip after synapsing with their appropriate target. Neuron 2:1063–1068

McGlade-McCulloh E, Morrissey AM, Norona F, Muller KJ (1989) Individual microglia move rapidly and directly to nerve lesions in the leech central nervous system. Proc Natl Acad Sci USA 86:1093–1097

McKerracher L, Hirscheimer A (1992) Slow transport of the cytoskeleton after axonal injury. J Neurobiol 23:568–578

McLane JA, Cobb T, Kahn T (1994) Electric field effects on axonal transport in the corticospinal tract following spinal injury. Soc Neurosci Abstr 20: 1324

McLean M, Edwards JS (1976) Target discrimination in regenerating insect sensory nerve. J Embryol Exp Morphol 36: 19–39

McQuarrie IG, Grafstein B, Gershon MB (1977) Axonal regeneration in the rat sciatic nerve. Effect of a conditioning lesion and of dbc-AMP. Brain Res 132:443–453

McQuarrie IG, Grafstein B (1982) Protein synthesis and axonal transport in goldfish retinae ganglion cells during regeneration accelerated by a conditioning lesion. Brain Res 251:25–37

McRorie JW, Zipser B (1988) Neurite outgrowth of identified neurons and the role of characterized surface glycoproteins in the leech. In: Beadle DJ, Lees G, Kater SB (eds) Cell culture approaches to invertebrate neuroscience. Academic Press, London, pp 33–52

McTeague JA, Govind CK (1993) Atrophy of nerve and neuropil following limb loss and regeneration in adult snapping shrimps, *Alpheus heterochelis*. J Exp Zool 265: 507–514

McVean AR (1975) Autotomy: minireview. Comp Biochem Physiol 51A: 497–506

Meadows LA, Gell D, Broadie K, Gould AP, White RHA (1994) The cell adhesion molecule, connectin, and the development of the *Drosophila* neuromuscular system. J Cell Science 107:321–328

Mearow KM, Kril Y, Gloster A, Diamond J (1994) Expression of NGF receptor and GAP-43 mRNA in DRG neurons during collateral sprouting and regeneration of dorsal cutaneous nerves. J Neurobiol 25:127–142

Meinertzhagen IA (1993) The synaptic populations of the fly's optic neuropil and their dynamic regulation: parallels with the vertebrate retina. Prog Retin Res 12:13–39

Meiri H, Grafstein B (1984) Local application of calcium-modulating agents to a crushed coldfish optic nerve modifies visual recovery. Exp Neurol 83:403–413

Meiri H, Spira ME, Parnas I (1981) Membrane conductance and action potential of a regenerating axonal tip. Science 211: 709–712
Meiri KF, Burdick D (1991) Nerve growth factor stimulation of GAP-43 phosphorylation in intact isolated growth cones. J Neurosci 11:3155–3164
Meiri KF, Gordon-Weeks PR (1990) GAP-43 in growth cones is associated with areas of membrane that are tightly bound to substrate and is a component of a membrane skeleton subcellular fraction. J Neurosci 10: 256–266
Merz DC, Drapeau P (1992) Cell-specific contact selects transmitter responses in an identified leech neuron. Proc R Soc Lond B Biol Sci 248:129–133
Merz DC, Drapeau P (1994) Cell surface contact mediates neuronal recognition and synapse formation between two identified leech neurons. J Neurobiol 25:1029–1037
Meyers CM, Whitington PM, Ball EE (1990) Embryonic development of the embryonic locust extensor tibiae muscle by identified neurons: formation and elimination of inappropriate axon branches. Dev Biol 137:194–206
Millar DA, Ratcliffe NA (1994) Invertebrates. In: Turner RJ (ed) Immunology: a comparative approach. John Wiley, Chichester, pp 29–68.
Miller DM, Shen MM, Shamu CE, Bürglin TR, Ruvkun G, Dubois ML, Ghee M, Wilson L (1992) *C. elegans unc-4* gene encodes a homeodomain protein that determines the pattern of synaptic input to specific motor neurons. Nature 355:841–845
Miller JD, Hadley RD (1991) Laminin-like immunoreactivity in the snail *Helisoma*: involvement of a ca. 300 kD extracellular matrix protein in promoting outgrowth from identified neurons. J Neurobiol 22: 431–442
Mills LR, Kater SB (1990) Neuron-specific and state specific differences in calcium homeostasis regulate the generation and degeneration of neuronal architecture. Neuron 2:149–163
Mimura K (1993) Effect of the visual environment on proteins and peptides in the developing brain of the fly, *Boettcherisca peregrina*. J Insect Physiol 39:145–151
Mittenthal J (1980) On the form and size of crayfish legs regenerated after grafting. Biol Bull 159:700–713
Moffett SB (1975) Motor patterns and structural interactions of basi-ischiopodite levator muscles in routine limb elevation and production of autotomy in the land crab *Cardisoma guanhumi*. J Comp Physiol 96:285–305
Moffett SB (1987) Muscles proximal to the fracture plane atrophy after limb autotomy in decapod crustaceans. J Exp Zool 244:485–490
Moffett SB (1990) Locomotion in the pulmonate snail *Melampus*. I. The motor pattern. Comp Biochem Physiol 96A:399–406
Moffett SB (1991a) Reinnervation controls regeneration of target organs in the snail *Melampus*. In: Kits KS, Boer HH, Joosse J (eds) Molluscan neurobiology. North Holland, Amsterdam, pp 168–173
Moffett SB (1991b) Neural control of male reproductive function in pulmonate and opisthobranch gastropods. In: Florey E, Stefano GB (eds) Comparative aspects of neuropeptide function. Manchester Univ Press, Manchester, pp 140–153
Moffett SB (1992) Mating behavior in the pulmonate snail *Melampus*: can regeneration restore function? Acta Biol Hung 43:369–376
Moffett SB (1993) Mating behavior in the pulmonate snail *Melampus*: can regeneration restore function? In: Salanki J, S-Rozsa K, Elekes K (eds) Neurobiology of invertebrates-signal molecules, networks, behavior. Akadémiai Kiado, Budapest, pp 367–374
Moffett SB (1995) Neural regeneration in gastropod molluscs. Prog Neurobiol 46:289–330
Moffett S, Austin DR (1982) Generation of new cerebral ganglion neurons in the snail *Melampus*: an ultrastructural study. J Comp Neurol 207:177–182
Moffett SB, Bale S (1993) Glial responses to nerve crush in a snail as revealed by neurotransmitter immunocytochemistry and the nuclear fluorochrome bisbenzimide. Soc Neurosci Abstr 19:45
Moffett SB, Ridgway RL (1988a) Central nervous system regeneration in the snail *Melampus bidentatus* as revealed by serotonin immunocytochemistry. Symp Biol Hung 36:213–229
Moffett SB, Ridgway RL (1988b) Structural repair and functional recovery following cerebral ganglion removal in the pulmonate snail *Melampus*. Am Zool 28:1109–1122

Moffett S, Snyder K (1985) Behavioral recovery associated with central nervous system regeneration in the snail *Melampus.* J Neurobiol 16:193–209
Moffett SB, Sidelnikov AP, Stepanov II (1995) Humoral control of mitotic activity in tentacle regeneration of the snail Achatina fulica. Soc Neurosci Abstr 21 (in press)
Moment GB (1951) Simultaneous anterior and posterior regeneration and other growth phenomena in maldanid polychaetes. J Exp Zool 117:1–14
Montell DJ, Goodman CS (1988) *Drosophila* substrate adhesion molecule: sequence of laminin B1 chain reveals domains of homology with mouse. Cell 53:463–473
Moody WJ, Simoncini L, Coombs JL, Spruce AE, Villaz M (1991) Development of ion channels in early embryos. J Neurobiol 22:674–684
Morgese VJ, Elliott EJ, Muller KJ (1983) Microglial movement to sites of nerve lesion in the leech CNS. Brain Res 272:166–170
Muller KJ (1988) Synapse regeneration, axon sprouting and degeneration in the central nervous system of the leech. Am Zool 28:1091–1097
Muller KJ, Carbonetto S (1979) the morphological and physiological properties of a regenerating synapse in the C.N.S. of the leech. J Comp Neurol 185:485–516
Muller KJ, Gu X (1991) Unequal competition between axons for neuronal targets. Philos Trans R Soc Lond B Biol Sci 331:315–322
Muller KJ, Scott SA (1979) Correct axonal regeneration after target cell removal in the central nervous system of the leech. Science 206:87–89
Muller KJ, Scott SA (1980) Removal of the synaptic target permits electrical sprouting of a mature intact axon. Nature 283:89–90
Muller KJ, Nicholls JG, Stent GS (1981) Neurobiology of the leech. Cold Spring Harbor Lab, New York, 320 pp
Muller KJ, McGlade-McCulloh E, Mason A (1987) Tinkering with successful synapse regeneration in the leech: adding insult to injury. J Exp Biol 132:207–221
Müller WA (1991) Stimulation of head-specific nerve cell formation in *Hydra* by pulses of diacylglycerol. Dev Biol 147:460–463
Müller WA (1991) Stimulation of head-specific nerve cell formation in *Hydra* by pulses of diacylglycerol. Dev Biol 147:460–463
Muramoto A (1993) Axotomy-induced long-lasting firing in an identified crayfish motoneuron. Zool Sci 10:913–924
Murphey RK (1986) The myth of the inflexible invertebrate: competition and remodelling in the development of invertebrate nervous systems. J Neurobiol 17:585–591
Murphey RK, Matsumoto SG, Mendenhall B (1976) Recovery from deafferentation by cricket interneurons after reinnervation by their peripheral field. J Comp Neurol 169:335–346
Murphey RK, Jacklet A, Schuster L (1980) A topographic map of sensory cell terminal arborizations in the cricket CNS: correlation with birthday and position in a sensory array. J Comp Neurol 169: 53–64
Murphey RK, Bacon JP, Sakaguchi DS, Johnson SE (1983a) Transplantation of cricket sensory neurons to ectopic locations: arborizations and synaptic connections. J Neurosci 3:659–672
Murphey RK, Johnson SE, Sakaguchi DS (1983b) Anatomy and physioloby of supernumerary cercal afferents in crickets: implications for pattern formation. J Neurosci 3:312–325
Murphy AD, Kater SB (1980) Differential discrimination of appropriate pathways by regenerating identified neurons in *Helisoma.* J Comp Neurol 190: 395–403
Murphy AD, Hadley RD, Kater SB (1984) Axotomy-induced parallel increases in electrical and dye coupling between identified neurons of *Helisoma.* J Neurosci 3: 1422–1429
Murrain M, Murphy AD, Mills LR, Kater SB (1990) Neuron-specific modulation by serotonin of regenerative outgrowth and intracellular calcium within the CNS of *Helisoma trivolvis.* J Neurobiol 21: 611–618
Muskavitch MAT, Hoffmann FM (1990) Homologies of vertebrate growth factors in *Drosophila melanogaster* and other invertebrates. In: Nilson-Hamilton M (ed) Current topics in developmental biology, vol 24. Academic Press, San Diego, pp 289–301
Myers CM, Whitington PM, Ball EE (1990) Embryonic development of the innervation of the locust extensor tibiae muscle by identified neurons: formation and elimination of inappropriate axon branches. Dev Biol 137:194–206

Nagasawa H, Kataoka H, Isogai A, Suzuki S, Suzuki A, Mizoguchi Y, Fugiwara Y, Susuki At, Takahashi SY, Ishizaki A (1986) Amino acid sequence of the silkworm *Bombyx mori.* Proc Natl Acad Sci USA 83:5840–5843
Nardi JB, Vernon RA (1990) Topographical features of the substratum for growth of pioneering neurons in the *Manduca* wing disc. J Neurobiol 21:1189–1201
Needham AE (1952) Regeneration and wound-healing. John Wiley, New York
Needham AE (1965) Regeneration in the arthropoda and its endocrine control. In: Kiortsis V, Trampusch HAL (eds) Regeneration in animals and related problems. North Holland, Amsterdam, pp 283–323
Newland PL, Smith PJS, Howes EA (1993) Regenerating adult cockroach dorsal unpaired median neurones in vitro retain their in vivo membrane characteristics. J Exp Biol 179:323–329
Nicholls JG (1986) The search for connections: studies of regeneration in the nervous system of the leech. Magnes Lecture Series Vol II. Sinauer, Sunderland Massachusetts, 86 pp
Nicholls JG, Hernandez UG (1989) Growth and synapse formation by identified leech neurones in culture: a review. J Exp Physiol 74:965–973
Nicholls JG, Van Essen D (1974) The nervous system of the leech. Sci Am 230:38–48
Norbeck BA, Feng Y, Denburg JL (1992) Molecular gradients along the proximal-distal axis of embryonic insect legs: possible guidance cues of pioneer axon growth. Development 116: 467–479
Nose A, Mahajan VB, Goodman CS (1992) Connectin: a homophilic cell adhesion molecule expressed on a substrate of muscles and the motoneurons that innervate them in *Drosophila.* Cell 70:553–567
Nottebohm F (1981) A brain for all seasons: cyclical anatomical changes in song control nuclei of the canary brain. Science 214:1368–1370
Nottebohm F, Nottebohm ME (1978) Relationship between song repertoire and age in the canary (*Serinus canarius*). Z Tierpsychol 46:298–305
O'Connor TP, Duerr JS, Bentley D (1990) Pioneer growth cone steering decisions mediated by single filopodial contacts in situ. J Neurosci 10: 3935–3946
Okabe S, Hirokawa N (1991) Actin dynamics in growth cones. J Neurosci 11:1918–1929
Okamura Y, Takahashi K (1993) Neural induction suppresses early expression of the inward-rectifier K^+ channel in the ascidian blstopore. J Physiol 1993: 245–268
Okamura Y, Okado H, Takahashi, K (1993) The ascidian embryo as a prototype of vertebrate neurogenesis. BioEssays 15:723–730
Olive PJW (1974) Cellular aspects of regeneration hormone influence in *Nereis diversicolor.* J Embryol Exp Morphol 32:111–131
Page LR (1992a) New interpretation of a nudibranch central nervous system based on ultrastructural analysis of neurodevelopment in *Melibe leonina.* I. Cerebral and visceral loop ganglia. Biol Bull 182: 348–365
Page LR (1992b) New interpretation of a nudibranch central nervous system based on ultrastructural analysis of neurodevelopment in *Melibe leonina.* II. Pedal, pleural, and labial ganglia. Biol Bull 182: 366–381
Page LR (1993) Developmental analysis reveals labial and subradular ganglia and the primary framework of the nervous system in nudibranch gastropods. J Neurobiol 24: 1443–1459
Palka J (1993) Neuronal specificity and its development in the *Drosophila* wing disc and its derivatives. J Neurobiol 24:788–802
Palka J, Edwards JS (1974) The cerci and abdominal giant fibres of the house cricket, *Acheta domesticus.* II. Regeneration and effects of chronic deprivation. Proc R Soc Lond B Biol Sci 185:105–121
Palka J, Whitlock KE, Murray MA (1992) Guidepost cells. Curr Opinion Neurobiol 2:48–54
Palladini B, Medolago-Albani L, Gallo VP, Lauro GM, Diana G, Scorsini D, Alfei L, Margotta V (1988) The answer of planarina *Dugesia gonocephala* neurons to nerve growth factor. Cell Mol Biol 34:53–63
Pallas SL, Hoy RR (1986) Regeneration of normal afferent input does not eliminate aberrant synaptic connections of an identified auditory interneuron in the cricket, *Teleogryllus oceanicus.* J Comp Neurol 248:348–359

Palmberg I (1986) Cell migration and differentiation during wound healing and regeneration in *Microstomum lineare* (Turbellaria). Hydrobiologia 1332:181–188
Palmberg I, Reuter M (1990) Neuronal subsets in regenerating *Microstomum lineare*. Immunocytochemistry of FMRF/RF-amide and 5-HT. In: Gustafsson, MKS, M Reuter (eds) The early brain. Proc Symp Invertebrate neurobiology. Abo Academu Press, Abo, pp 147–160
Parente RA, Lentz BR (1986) Rate and extent of poly(ethylene glycol)-induced large vesicle fusion monitored by bilayer and internal contents mixing. Biochemistry 25:6678–6688
Parnas I, Dudel J, Atwood HL (1991) Synaptic transmission in decentralized axons of rock lobster. J Neurosci 11(5):1309–1315
Parnas I, Dudel J, Cohen I, Franke CH (1984) Strengthening of synaptic contacts of an excitatory axon on elimination of a second excitatory axon innervating the same target. J Neurosci 4:1912–1923
Parsons TD, Salzberg BM, Obaid AL, Raccuia-Behling F, Kleinfeld D (1991) Long-term optical recording of patterns of electrical activity in ensembles of cultured *Aplysia* neurons. J Neurophys 66:316–333
Passani MB, Peinado A, Engelman H, Baptista CA, Macagno ER (1991) Normally unused positional cues guide ectopic afferents in the leech CNS. J Neurosci 11:3868–3876
Patel NH, Snow PM, Goodman CS (1987) Characteristics and cloning of fasciclin III: a glycoprotein expressed on a subset of neurons and axon pathways in *Drosophila*. Cell 48:975–988
Patterson PH (1988) On the importance of being inhibited, or saying no to growth cones. Neuron 1:263–267
Pearson KG, Bradley AB (1972) Specific reinnervation of excitatory motoneurons to leg muscles in the cockroach. Brain Res 47:492–496
Peinado A, Zipser B, Macagno E (1987a) Regeneration of afferent axons into discrete tracts within peripheral nerves in the leech. Brain Res 410:330–334
Peinado A, Zipser B, Macagno E (1987b) A group of related surface glycoproteins distinguish sets and subsets of sensory afferents in the leech nervous system. Brain Res 410:335–339
Pellegrino M, Nencioni B, Matteoli M (1984) Response to axotomy of an identified leech neuron, in vivo and in culture. Brain Res 298:347–352
Perry VH, Andersson P-B, Gordon S (1993) Macrophages and inflammation in the central nervous system. Trends Neurosci 16:268–273
Pestarino M (1993) Localization of peptide histidine isoleucine-like immunoreactivity in the gut of the ascidian, *Styela plicata*. Anim Biol 2:91–95
Pestarino M, Lucaroni B, Distefano S (1993) Presence and distribution of immunoreactive FMRFamide- and bovine pancreatic polypeptide-like peptides in a protochordate ascidian. Eur J Histochem 37:225–232
Pfannenstiel H-D (1984) The ventral nerve cord signals positional information during segment formation in an annelid (*Ophryotrocha puerilis* Polychaeta). Roux's Arch Dev Biol 194:32–36
Pfenninger KH de la, Houssaye BA, Frame L, Helmke S, Lockerbie RO, Lohse K, Miller V, Neger-Aminou P, Wood MR (1990) Biochemical dissection of plasmalemmal expansion at the growth cone. In: Letourneau PC, Kater SB, Macagno ER (eds) The nerve growth cone. Raven Press, New York, pp 111–123
Pilkington JB, Benjamin PR (1986) Voltage clamp-measured decrease in junctional resistance followed extrajunctional axotomy of a pair of electrotonically coupled snail neurons. Proc R Soc Lond B Biol Sci 228:523–526
Pinter MJ, Noven SV, Muccio D, Wallace N (1991) Axotomy-like changes in cat motoneuron electrical properties elicited by botulinum toxin depend on the complete elimination of neuromuscular transmission. J Neurosci 11:657–666
Piovant M, Léna P (1988) Membrane glycoproteins immunologically related to the human insulin receptor are associated with presumptive neuronal territories and developing neuroines in *Drosophila melanogaster*. Development 103:145–156
Pitman RM, Rand KA (1982) Neural lesions can cause dendritic sprouting of an undamaged adult insect motoneuron. J Exp Biol 96:125–130
Pitman RM, Tweedle CD, Cohen MJ (1972) Electrical responses of insect central neurons: augmentation by nerve section or colchicine. Science 178:507–509

Plickert G (1989) Proportion-altering factor (PAF) stimulates nerve cell formation in *Hydractinia echinata*. Cell Differ Dev 26:19–28
Polak KA, Edelman AM, Wasley JWF, Cohan CS (1991) A novel calmodulin antagonist, CGS 9343B, modulates calcium-dependent changes in neurite outgrowth and growth cone movements. J Neurosci 11: 534–542
Price CH (1977) Regeneration in the central nervous system of a pulmonate snail *Melampus*. Cell Tissue Res 180:529–536
Purves D, Snider WD, Voyvodic JT (1988) Trophic regulation of nerve cell morphology and innervation in the autonomic nervous system. Nature 336:123–128
Rankin CH, Carew TJ (1987) Development of learning and memory in *Aplysia*. II. Habituation and dishabituation. J Neurosci 7:133–143
Raper JA, Bastiani M, Goodman CS (1983) Pathfinding by neuronal growth cones in grasshopper embryos. I. Divergent choices made by the growth cones of sibling neurons. J Neurosci 3:20–30
Reber-Muller S, Ono S, Wehrle-Haller M, Schmid V (1994) Transdifferentiation of striated muscle of jellyfish to smooth muscle and nerve cells: the role of cell-ECM interactions and carbohydrates revealed by monoclonal antibody. Differentiation 57:77–87
Rehder V, Kater SB (1992) Regulation of neuronal growth cone filopodia by intracellular calcium. J Neurosci 12:3175–3186
Reichardt LF, Tomaselli KJ (1991) Extracellular matrix molecules and their receptors: functions in neural development. Annu Rev Neurosci 14:531–570
Reichert H (1993) Molecular correlates of neuronal specificity in the developing insect nervous system. Mol Neurobiol 7:349–362
Reinecke M, Betzler D, Drakenberg K, Falkmer S, Sara VR (1993) Occurrence of members of the insulin superfamily in central nervous system and digestive tract of protochordates. Histochemistry 99: 277–285
Reis DJ, Ross RA (1973) Dynamic changes in dopamine-beta-hydroxylase activity during the anterograde and retrograde reaction to injury of central noradrenergic axons. Brain Res 57:307–326
Reuter M, Palmberg I (1989) Development and differentiation of neuronal subsets in asexually reproducing *Microstomum lineare*. Histochemistry 91:123–131
Reuter M, Kuusisto A (1992) Growth factors in asexually reproducing Catenulida and Macrostomida (Plathelminthes)? Zoomorphology 112:155–166
Reynolds BA, Weiss S (1992) Generation of neurons and astro
cytes from isolated cells of the adult mammalian central nervous system. Science 255:1707–1710
Richmond JE, Funte LR, Smith WL, Price DA, Haydon PG (1991) Activation of a peptidergic synapse modulates postsynaptic calcium influx. J Exp Biol 161: 257–271
Richards J, Pasco D, Yang J, Guzman R, Nandi S (1983) Comparison of the growth of normal and neoplastic mouse mammary cells on plastic, on collagen gels and in collagen gels. Exp Cell Res 146:1–14
Ridgway RL, Bulloch AGM (1991) Glutamate modulation of sprouting by adult *Helisoma* neurons. In: Kits KS, Boer HH, Joosse J (eds) Molluscan neurobiology. North Holland, Amsterdam pp154–160
Ridgway RL, Syed NI, Bulloch AGM (1989) Neurite outgrowth of *Helisoma* and *Lymnaea* neurons is enhanced by in vitro co-culture with endocrine dorsal body cells. Eur J Neurosci [Suppl] 2:143
Ridgway RL, Syed NI, Lukowiak K, Bulloch AGM (1991) Nerve growth factor (NGF) induces sprouting of specific neurons of the snail, *Lymnaea stagnalis*. J Neurobiol 22:377–390
Roberts A, Mackie GO (1980) The giant axon escape system of a hydrozoan medusa, *Aglantha digitale*. J Exp Biol 84:303–318
Roederer E, Cohen MJ (1983) Regeneration of an identified central neuron in the cricket. II. Electrical and morphological responses of the soma. J Neurosci 3:1848–1859
Roger I, Syed NI, Ridgway RL, Lukowiak K, Bulloch AGM (1992) Regeneration of an interneuronal network in *Helisoma*: re-establishment of synaptic contacts. Acta Biol Hung 43(1–4): 99–111

Ross RA, Joh ML, Reis DJ (1975) Reversible changes in the accumulation and activities of tyrosine hydroxylase and dopamine beta-hydroxylase in neurons of the locus ceruleus during the retrograde reaction. Brain Res 92:57–72

Ross TL, Govind CK, Kirk MD (1994) Neuromuscular regeneration by buccal motoneuron B15 after peripheral nerve crush in Aplysia californica. J Neurophysiol 72:1897–1910

Ross WN, Arechiga H, Nicholls JG (1987) Optical recording of calcium and voltage transients following impulses in cell bodies and processes of identified leech neurons in culture. J Neurosci 7:3877–3887

Ross WN, Arechiga H, Nicholls JG (1988) Influence of substrate on the distribution of calcium channels in identified neurons in culture. Proc Natl Acad Sci USA 85:4075–4078

Rossiter JP, Riopelle RJ, Bisby MA (1994) Bromodeoxyuridine incorporation by axotomized neonatal rat facial motoneurons undergoing apoptotic cell death. Soc Neurosci Abstr 20:245

Rotshenker S (1979) Synapse formation in intact innervated cutaneous-pectoris muscles of the frog following denervation of the opposite muscle. J Physiol (Lond) 292:535–547

Roubos EW, van Winkoop A, van der Haar C, van Minnen J (1988) Postembryonic development of endocrine dorsal bodies and neuroendocrine egg laying and growth hromone producing neurons of *Lymnaea stagnalis.* Int J Invertebr Reprod Dev 13:119–145

Rubel EW, Dew LA, Roberson DW (1995) Mammalian vestibular hair cell regeneration. Science 267:701–703

Rutishauser U, Acheson A, Hall AK, Mann DM, Sunshine J (1988) The neural cell adhesion molecule (NCAM) as a regulator of cell-cell interactions. Science 240:53–57

Ryerse JS (1989) Isolation and characterization of gap junctions from *Drosophila melanogaster.* Cell Tissue Res 256:7–16

Sabry JH, O'Connor TP, Evans L, Toroian-Raymond A, Kirshner M, Bentley D (1991) Microtubule behavior during guidance of pioneer neuron growth cones in situ. J Cell Biol 115:381–395

Saló E, Baguña J (1986) Stimulation of cellular proliferation and differentiation in the intact and regenerating planarian *Dugesia(G) tigrina* by the neuropeptide substance P. J Exp Zool 237: 129–135

Saló E, Baguña J (1989) Regeneration and pattern formation in planarians II. Local origin and role of cell movements in blastema formation. Development 107:69–76

Sanes JR (1989) Extracellular matrix molecules that influence neural development. Annu Rev Neurosci 12:491–516

Sawyer RT (1981) Leech biology and behavior. In: Muller KJ, Nicholls JG, Stent GS (eds) Neurobiology of the leech. Cold Spring Harbor Lab, New York, pp 7–26

Schacher S, Kandel ER, Woolley R (1979) Development of neurons in the abdominal ganglion of *Aplysia californica.* 1. Axosomatic synaptic contacts. Dev Biol 71:163–175

Schachner M (1991) Neural recognition molecules and their influence on cellular functions. In: Letourneau PC, Kater SB, Macagno ER (eds) The nerve growth cone. Raven Press, New York, pp 237–254

Schaller HC, Hoffmeister AH, Dubel S (1989) Role of neuropeptide head activator for growth and development in hydra and mammals. Development (1989 Suppl):99–107

Schaller HC, Hofmann M, Javois LC (1990) Effect of head activator on proliferation, head-specific determination and differentiatioon of epithelial cells in hydra. Differentiation 43:157–164

Schmiege DL, Ridgway RL, Moffett SB (1992) Ultrastructure of autotomy-induced atrophy of muscles in the crab *Carcinus maenas.* Can J Zool 70: 841–851

Schildberger K, Wohlers DW, Schmitz B, Kleindienst HU, Huber F (1986) Morphological and physiological changes in central auditory neurons following unilateral foreleg amputation in larval crickets. J Comp Physiol 158A:291–300

Schmitz B (1989) Neuroplasticity and phonotaxis in monaural adult female crickets (*Gryllus bimaculatus* deGeer). J Comp Physiol 164A:343–358

Schmitz B, Kleindienst HU, Schildberger K, Huber F (1988) Acoustic orientation in adult, female crickets (*Gryllus bimaculatus* deGeer) after unilateral foreleg amputation in the larva. J Comp Physiol 162A:715–728

Schuch U, Lohse MJ, Schachner M (1989) Neural cell adhesion molecules influence second messenger systems. Neuron 3:13–20
Schultze LS (1899) Die Regeneration des Ganglions von *Ciona intestinalis* L. und über das Verhältnis der Regeneration und Knospung zur Keimenblätterlehre. Naturwissenschaft (Jena) 33:263–344
Schwartz M (1987) Molecular and cellular aspects of nerve regeneration. CRC Crit Rev Biochem 22, (2):89–110
Scott ML, Kirk MD (1992) Recovery of consummatory feeding behavior after bilateral lesions of the cerebral-buccal connectives in *Aplysia californica*. Brain Res 585:272–274
Seaver EC, Karlstrom RO, Bastiani MJ (1991) The restricted temporal and spatial expression of a nervous-system-specific antigen involved in axon outgrowth during development of the grasshopper. Development 111:881–893
Seshan KR, Bittner GD (1987) Developmental and other factors affecting regeneration of crayfish CNS axons. J Comp Neurol 262:535–545
Seshan KR, Levi-Montalcini R (1973) Neuronal properties of nymphal and adult insect neurowecretory cells in vitro. Science 182:291–293
Sheller RA, Ballinger ML, Bittner GD (1991) Long-term survival of severed crayfish giant axons is not associated with an incorporation of glial nuclei into the axoplasm. Neurosci Lett 133:113–116
Sheng M, Greenberg ME (1991) The regulation and function of c-fos and other immediate early genes in the nervous system. Neuron 4:477–485
Sheng M, Thompson MA, Greenberg ME (1991) CREB: a Ca^{++}-regulated transcription factor phosphorylated by calmodulin-dependent kinases. Science 252: 1427–1430
Shepherd D, Murphey RK (1986) Competition regulates the efficacy of an identified synapse in crickets. J Neurosci 6:3152–3160
Shepherd D, Kamper G, Murphey RK (1988) The synaptic origins of receptive field properties in the cricket cercal sensory system. J Comp Physiol (A) 162:1–11
Sidelnikov AP (1991) Optic tentacle regeneration in giant African snail *Achatina fulica*. Proc Acad Sci USSR, Ser Biol 18:716:725 (in Russian)
Sidelnikov AP, Stepanov II (1992) The influence of intact and regenerating snail's hemolymph on the value of mitotic index of snail recipients. Ontogenez 23(N3):313 (in Russian)
Sigurdson WJ, Morris CE (1989) Stretch-activated ion channels in growth cones of snail neurons. J Neurosci 9: 2801–2808
Silver AL, Lamb AG, Bolsover SR (1990) Calcium hotspots caused by L-channel clustering promote morphological changes in neuronal growth cones. Nature 343:751–754
Singer M (1959) The influence of nerves on regeneration. In: Thornton CS(ed) Regeneration in vertebrates. Univ Chicago Press, Chicago, pp 59–80
Singer M, Nordlander RH, Edgar M (1979) Axonal guidance during embryogenesis and regeneration in the spinal cord of the newt: the blueprint hypothesis of neural pathway patterning. J Comp Neurol 185:1–22
Singla CL (1978) Locomotion and neuromuscular system of *Aglantha digitale*. Cell Tissue Res 188:317–327
Sink H, Whitington PM (1991a) Early ablation of target muscles modulates the arborisation pattern of an identified embryonic *Drosophila* motor axon. Development 113:701–707
Sink H, Whitington PM (1991b) Pathfinding in the central nervous system and periphery by identified embryonic *Drosophila* motor neurons. Development 112:307–316
Skene JHP (1992) Retrograde pathways controlling the expression of a major growth cone component in the adult CNS. In: Letourneau PC, Kater SB, Macagno ER (eds) The nerve growth cone. Raven Press, New York, pp 463–475
Skinner DM (1985) Molting and regeneration. In: Bliss DE, Mantel LG (eds) Integuments, pigments and hormonal processes. The biology of Crustacea. vol 9. Academic Press, New York, pp 43–146.
Skinner DM, Graham DE (1972) Loss of limbs as a stimulus to ecdysis in Brachyura (true crabs). Biol Bull 143:222–233
Smalheiser NR, Crain SM (1984) The possible role of „sibling neurite bias" in the coordination of neurite extension, branching and survival. J Neurobiol 15:517–529

Smith PJS, Howes EA (1987) Neural repair in an insect central nervous system: cell kinetics and proliferation after selective glial disruption. Cell Tissue Res 247:129–135
Smith PJS, Howes EA, Treherne JE (1990) Cell proliferation in the repairing adult insect central nervous system: incorporation of the thymidine analogue 5-bromo-deoxyuridine in vivo. J Cell Sci 95:599–604
Smith PJS, Shepherd D, Edwards JS (1991) Neural repair and glial proliferation: parallels with gliogenesis in insects. BioEssays 13:65–72
Smith SJ, Augustine GJ (1988) Calcium ions, active zones and synaptic transmitter release. Trends Neurosci 11:458–464
Snider WD, Johnson EM (1989) Neurotrophic molecules. Ann Neurol 26:489–506
Snyder K, Moffett SB (1987) Loss and recovery of locomotory behavior after CNS lesions in the snail *Melampus bidentatus.* Biol Bull 172:83–88
Snyder K, Moffett SB (1990) Locomotion in the pulmonate snail *Melampus.* II. Recovery after pedal ganglion excision. Comp Biochem Physiol 96A:407–414
Sonetti D, Bianchi F (1993) Occurrence and distribution of insulin receptor-like immunoreactivity in molluscan brains. Acta Biol Hung 44:77–82
Sonetti D, van Heumen WRA, Roubos EW (1992) Light- and electron-microscopic immunocytochemistry of a molluscan insulin-related peptide in the central nervous system of *Planorbarius corneus.* Cell Tissue Res 267:473–481
Spencer AN, Satterlie RA (1981) The action potential and contraction in subumbrellar swimming muscle of *Polyorchis penicillatus* (Hydromedusae). J Comp Physiol 144:401–407
Sperry PJ, Ansevin KD, Tittel FK (1973) Inductive role of the nerve cord in regeneration of isolated postpharyngeal body sections of *Dugesia dorotocephala.* J Exp Zool 186:159–174
Spira ME, Zeldes D, Hochner B, Dormann A (1987) The effects of microenvironment on the redifferentiation of regenerating neurones: neurite architecture, acetylcholine receptors and Ca^{2+} channel distribution. J Exp Biol 132:111–131
Spira ME, Benbassat D, Dormann A (1993) Resealing of the proximal and distal cut ends of transected axons: electrophysiological and ultrastructural analysis. J Neurobiol 24:300–316
Spitzer NC (1991) A developmental handshake: neuronal control of ionic currents and their control of neuronal differentiation. J Neurobiol 22:659–673
Stasek CR (1967) Autotomy in the Mollusca. Occas Pap Calif Acad Sci 61:1–44
Stefano GB (1992) Invertebrate and vertebrate neuroimmune and autoimmunoregulatory commonalities involving opioid peptides. Cell Mol Neurobiol 12:357–366
Steinhardt RA, Bi G, Alderton JM (1994) Cell membrane resealing by a vesicular mechanism similar to neurotransmitter release. Science 263:390–393
Stent GS (1973) A physiological postulate for Hebb's postulate of learning. Proc Natl Acad Sci USA 70:997–1001
Stent GS, Kristin WB Jr, Torrence SA, French KA, Weisblat DA (1992) Development of the leech nervous system. Int Rev Neurobiol 33:109–193
Stewart BA, Atwood HL (1992) Synaptic plasticity in a regenerated crayfish phasic motoneuron. J Neurobiol 23:881–889
Stewart R, McCaig CD (1994) Calcium entry and calcium stores control nerve growth in an applied electric field. Soc Neurosci Abstr 20:1294
Strandburg RJ, Krasne FB (1985) Intracellular analysis of the innervation of a crayfish interneuron by regenerating afferents. J Neurophysiol 54:385–402
Strautman AF, Cork RJ, Robinson KR (1990) The distribution of free calcium in transected spinal axons and its modulation by applied electrical fields. J Neurosci 10:3564–3575
Strittmatter SM, Fishman MC (1991) The neuronal growth cone as a specialized transduction system. Bioessays 13:127–134
Stuart DK, Blair SS, Weisblat DA (1987) Cell lineage, cell death, and the development origin of identified serotonin- and dopamine-containing neurons in the leech. J Neurosci 7(4):1107–1122
Stuermer CAO, Bastmeyer M, Bähr M, Strobel G, Paschke K (1992) Trying to understand axonal regeneration in the CNS of a fish. J Neurobiol 23:537–550
Syed NI, Bulloch AGM, Lukowiak K (1990) In vitro reconstruction of the respiratory central pattern generator of the mollusk *Lymnaea.* Science 250:282–285

Syed NI, Ridgway RL, Lukowiak K, Bulloch AGM (1992a) Transplantation and functional integration of an identified respiratory interneuron in *Lymnaea stagnalis*. Neuron 8:767–774
Syed NI, Lukowiak K, Bulloch AGM (1992b) Specific *in vitro* synaptogenesis between identified *Lymnaea* and *Helisoma* neurons. NeuroReport 3:793–796
Syed NI, Roger I, Ridgway RL, Bauce LG, Lukowiak K, Bulloch AGM (1993) Identification, characterisation and in vitro reconstruction of an interneuronal network of the snail *Helisoma trivolvis*. J Exp Biol 174:19–44
Taghert PH, Bastiani MJ, Ho RK, Goodman CS (1982) Guidance of pioneer growth cones: filopodial coupling and contacts revealed with an antibody to Lucifer yellow. Dev Biol 94:391–399
Thoenen H (1991) The changing scene of neurotrophic factors. Trends Neurosci 14:165–170
Thomson AM (1990) Epigenesis and plasticity in the nervous system. Trends Neurosci 13:389–391
Thorey IS, Zipser B (1991) The segmentation of the leech nervous system is prefigured by myogenic cells at the embryonic midline expressing a muscle-specific matrix protein. J Neurosci 11:1786–1799
Titmus MJ, Faber DS (1990) Axotomy-induced alterations in the electrophysiological characteristics of neurons. Prog Neurobiol 35:1–51
Tiwari SK, Woodruff ML (1992) *Helix aspersa* neurons maintain vigorous electrical activity when co-cultured with intact *H. aspersa* ganglia. Comp Biochem Physiol 101:163–174
Todorov AT, Yogev D, Qi P, Fendler JH, Rodziewicz GS (1992) Electric-field-induced reconnection of severed axons. Brain Res 582:329–334
Tolbert LP, Oland LA (1990) Glial cells form boundaries for developing insect olfactory glomeruli. Exp Neurol 109:19–28
Tomlinson A, Ready DF (1987) Neuronal differentiation in the *Drosophila* ommatidium. Dev Biol 120:336–366
Tornqvist E, Aldskogius H, Johansson K, Svennson M (1994) Disturbances in axonal transport trigger the neuronal signal(s) which activate microglia. Soc Neurosci Abstr 20:874
Torrence SA, Law MI, Stuart DK (1989) Leech Neurogenesis. II. Mesodermal control of neuronal patterns. Dev Biol 136:40–60
Tosney KW, Landmesser LT (1985) Growth cone morphology and trajectory in the lumbosacral region of the chick embryo. J Neurosci 5:2345–2358
Travis J (1992) Can 'hair cells' unlock deafness? Science 257:1344–1345
Travis J (1994) Wiring the nervous system. Science 266:568–570
Treherne JE, Smith PJS, Edwards H (1987) Neural repair in an insect: cell recruitment and deployment following selective glial disruption. Cell Tissue Res 247:121–128
Treherne JE, Smith PJS, Howes EA (1988) Cell recruitment during glial repair: the role of exogenous cells. Cell Tissue Res 251:339–343
Tribut F, Lapied B, Duval A, Pelhate M (1991) A neosynthesis of sodium channels is involved in the evolution of the sodium current in isolated adult DUM neurons. Pfluegers Arch Europ J Physiol 419:665–667
Truman JW (1990) Metamorphosis of the central nervous system of *Drosophila*. J Neurobiol 21:1072–1084
Tublitz NJ (1993) Steriod-induced transmitter plasticity in insect peptidergic neurons. Comp Biochem Physiol 105(C):147–154
Tucker M (1959) Inhibitory control of regeneration in nemertean worms. J Morphol 105:569–600
Turrigiano G, Abbott LF, Marder E (1994) Activity-dependent changes in the instrinsic properties of cultured neurons. Science 264:974–977
Uemura H, Tezuka Y, Hasegawa C, Kobayashi H (1994) Immunohistochemical investigation of neuropeptides in the central nervous system of the amphioxus, *Branchiostoma belcheri*. Cell Tissue Res 277:279–287
Ushkaryov YA, Petrenko AG, Geppert M, Sudhof TC (1992) Neurexins: Synaptic cell surface proteins related to the a-latrotoxin receptor and laminin. Science 257:50–56
Van Berkum MFA, Goodman CS (1995) Targeted disruption of Ca^{2+} -calmodulin signaling in *Drosophila* growth cones leads to stalls in axon extension and errors in axon guidance. Neuron 14:43–56

Van Essen DC, Jansen JKS (1977) The specificity of reinnervation by identified sensory and motor neurons on the leech. J Comp Neurol 171:433–454
Vanhems E, Delbos M (1989) Ultrastructural localization of lectin-binding sites and laminin-like immunoreactivity in glial cells and neurites growing out from explant cultures of the central nervous system of embryonic locusts. Cell Tissue Res 258:429–436
Vanhems E, Delbos M, Girardie J (1990) Insulin and neuroparsin promote neurite outgrowth in cultured locust CNS. Eur J Neurosci 2:776–782
Vanhems E, Delbos M, Girardie J (1993) In vitro growth of locust embryonic pars intercerebralis neurosecretory cells. Neurosci 52:361–368
Van Vactor D, Sink H, Fambrough D, Tsoo R, Goodman CS (1993) Genes that control neuromuscular specificity in *Drosophila*. Cell 73:1137–1153
Venstrom KA, Reichardt LF (1993) Extracellular matrix. 2. Role of extracellular matrix molecules and their receptors in the nervous system. FASEB J 7:996–1003
Viancour TA, Bittner GD, Seshan Ks, Sheller RA (1988) Protein transport between crayfish lateral giant axons. Brain Res 439:211–221
Vining EP, Drewes CD (1985a) Donor-recipient interconnections between giant nerve fibers in transplanted ventral nerve cords of earthworms. J Neurobiol 16:283–299
Vining EP, Drewes CD (1985b) Restoration of sensory and motor function in earthworm escape reflex pathways following ventral nerve cord transplantation. J Neurobiol 16:301–315
Vyklicky L, Nicholls J (1988) Specificity of connections formed by nociceptive cells of the leech in tissue culture. J Exp Biol 134:17–26
Wall JB, Taghert PH (1991) The timing of initial expression by an identified insect neuron does not depend upon interactions with its normal peripheral target. J Neurobiol 22:935–956
Walters ET (1987a) Site-specific sensitization of defensive reflexes in *Aplysia*: a simple model of long-term hyperalgesia. J Neurosci 7:400–407
Walters ET (1987b) Multiple sensory neuronal correlates of site-specific sensitization in *Aplysia*. J Neurosci 7:408–417
Walters ET (1991) A functional, cellular, and evolutionary model of nociceptive plasticity in *Aplysia*. Biol Bull 180:241–251
Walters ET, Ambron RT (1995) Long-term alterations induced by injury and by 5-HT in *Aplysia* sensory neurons: convergent pathways and common signals? Trends Neurosci 18:137–142
Walters ET, Alizadeh H, Castro GA (1991) Similar neuronal alterations induced by axonal injury and learning in *Aplysia*. Science 253:797–799
Walthall WW, Chalfie M (1988) Cell-cell interactions in the guidance of late-developing neurons in *Caenorhabditis elegans*. Science 239:643–645
Walthall WW, Li L, Plunkett JA, Hsu C-Y (1993) Changing synaptic specificities in the nervous system of *Caenorhabditis elegans*: differentiation of the DD motoneurons. J Neurobiol 12:1589–1599
Wang L, Feng Y, Denburg JL (1992) A multifunctional cell surface developmental stage-specific antigen in the cockroach embryo: involvement in pathfinding by CNS pioneer axons. J Cell Biol 118:163–176
Webster G, Hamilton S (1972) Budding in *Hydra* – Role of cell multiplication and cell movement in bud initiation. J Emb Exp Morph 27:301–316
Weeks JC (1990) Postembryonic neuronal plasticity and its hormonal control during insect metamorphosis. Annu Rev Neurosci 13:183–194
Weeks JC, Truman JW (1986) Hormonally mediated reprogramming of muscles and motoneurons during the larval-pupal transformation of the tobacco hornworm, *Manduca sexta*. J Exp Biol 125:1–13
Weeks JC, Ernst-Utzschneider K (1989) Respecification of larval proleg motoneurons during metamorphosis of the tobacco hornworm, *Manduca sexta*: segmental dependence and hormonal regulation. J Neurobiol 20:569–592
Weeks JC, Roberts WM, Trimble DL (1992) Hormonal regulation and segmental specificity of motoneuron phenotype during metamorphosis of the tobacco hornworm *Manduca sexta*. Dev Biol 149:185–196
Wehrle B, Chiquet M (1990) Tenascin is accumulated along developing peripheral nerves and allows neurite outgrowth in vitro. Development 110:401–415

Weiser M, Baker H, Wessel TC, Joh TH (1993) Differential spatial and temporal gene expression in response to axotomy and deafferentation following transection of the medial forebrain bundle. J Neurosci 13:3472–3484
Westfall JA, Saripalli LD, Sayyar KL (1994) Scanning electron microscopy of neurons dissociated from tentacles of a sea anemone. Soc Neurosci Abstr 20:775
Westin J, Camhi JM (1975) Regeneration of peripheral nerves to supernumerary legs in cockroaches. J Exp Biol 63:483–495
White JG, Southgate E, Thomson JN (1991) On the nature of the undead cells in the nematode *Caenorhabditis elegans*. Phil Trans R Soc Lond B 331: 263–271
White JG, Southgate E, Thomson JN (1992) Mutations in the *Caenorhabditis elegans unc-4* gene alter the synaptic input to ventral cord motor neurons. Nature 355:838–841
White JG, Southgate E, Thomson JN, Brenner S (1986) The structure of the nervous system of *Caenorhabditis elegans*. Philos Trans R Soc Lond B Biol Sci 314:1–340
Whitington PM (1979) The specificity of innervation of regenerating motor neurons in the cockroach. J Comp Neurol 186:465–472
Whitington PM (1993) Axon guidance factors in invertebrate development. Pharmacol Ther 58:263–299
Wilkens JL (1994) Degeneration of afferent neurons and long-term stability of the ventilatory central pattern generator in chronocially deafferented crabs. J Comp Physiol A 174:211–220
Wilson RJA, Blackshaw SE, Dow JAT (1991) Topographical discontinuities modify the outgrowth of leech Retzuis neurones grown in culture. J Physiol 438:138P
Wittenberg G, Loer CM, Adamo SA, Kristan WB Jr (1990) Segmental specialization of neuronal connectivity in the leech. J Comp Physiol A 1657:453–439
Wojtowitz JM, Marin L, Atwood HL (1989) Synaptic restructuring during long-term facilitation at the crayfish neuromuscular junction. Can J Physiol Pharmacol 67:167–171
Wojtowicz JM, Smith BR, Atwood HL (1991) Activity-dependent recruitment of silent synapses. Ann New York Acad Sci 627:169–179
Wong RG, Hadley RD, Kater SB, Hauser GC (1981) Neurite outgrowth in molluscan organ and cell cultures: the role of conditioning factor(s). J Neurosci 1:1008–1021
Worden MK, Hwang J-C, Vélez SJ (1988) Regeneration studies on a crayfish neuromuscular system. I. Connectivity changes after intersegmental nerve transplants. J Neurobiol 19:127–140
Wu CF (1988) Neurogenetic studies of *Drosophila* central nervous system neurons in culture. In: Beadle DJ, Lees G, Kater SB (eds) Cell culture approaches to invertebrate neuroscience. Academic Press, New York, pp 149–187
Wu C-F, Suzuki N, Poo M-M (1983) Dissociated neurons from normal and mutant *Drosophila* larval central nervous systems in cell culture. J Neurosci 3:1888–1899
Xie X, Barrett JN (1991) Membrane resealing in cultured rat septal neurons after neurite transection: evidence for enhancement by Ca^{2+}-triggered protease activity and cytoskeletal disassembly. J Neurosci 11:3257–3267
Xue Z, Grossfeld RM (1993) Stress protein synthesis and accumulation after traumatic injury of crayfish CNS. Neurochem Res 18:209–218
Yaross MS, Westerfield J, Javois LC, Bode HR (1986) Nerve cells in hydra: monoclonal antibodies identify two lineages with distinct mechanisms for their incorporation into head tissue. Dev Biol 114:225–237
Yawo H, Kuno M (1983) How a nerve fiber repairs its cut end: involvement of phospholipase A_2. Science 222:1351–1353
Yawo H, Kuno M (1985) Calcium dependence of membrane sealing at the cut end of the cockroach giant axon. J Neurosci 5:1626–1632
Yogev D, Todorov AT, Qi P, Fendler JH, Rodziewicz GS (1991) Laser-induced reconnection of severed axons. Biochem Biophys Res Comm 180:874–880
Young D (1972) Specific reinnervation of limbs transplanted between segments in the cockroach, *Periplaneta americana*. J Exp Biol 57: 305–316
Young D (1973) Specificity and regeneration in insect motor neurons. In: Young, D (ed) Developmental neurobiology of arthropods. Cambridge Univ Press, London, pp 179–202
Young JZ (1932) On the cytology of the neurons of cephalopods. Q J Microsc Sci 75:1–49

Zhang R-J, Nicholls J (1983) Identified neurons implanted into leech ganglia sprout and form synaptic connections. Dev Brain Res 8:131–135
Zhu H, Wu F, Schacher S (1994) *Aplysia* cell adhesion molecules and serotonin regulate sensory cell-motor cell interactions during early stages of synapse formation in vitro. J Neurosci 14:6886–6900
Ziller-Sengel C (1965) Inhibition de la régénération du pharynx chez les planaires. In: Kiortsis V, Trampusch HAL (eds) Regeneration in animals and related problems. North Holland, Amsterdam
Zimmerman U (1986) Electrical breakdown, electropermeabilization, and electrofusion. Rev Biochem Pharmacol 105:175–256
Zipser B, Cole RN (1991) A mannose-specific recognition mediates the defasciculation of axons in the leech CNS. J Neurosci 11:3471–3480
Zipser B, Morell R, Bajt ML (1989) Defasciculation as a neuronal pathfinding strategy: involvement of a specific glycoprotein. Neuron 3:621–630
Zipser K, Erhardt M, Song J, Cole RN, Zipser B (1994) Distributin of carbohydrate epitopes among disjoint subsets of leech sensory afferent neurons. J Neurosci 14(7):4481–4493
Zoran MJ, Drewes CD (1987) Rapid escape reflexes in aquatic oligochaetes: variations in design and function of evolutionarily conserved giant fiber systems. J Comp Physiol 161:729–738
Zoran MJ, Doyle RT, Haydon PG (1990) Target-dependent induction of secretory capabilities in an identified motoneuron during synaptogenesis. Dev Biol 138:202–213
Zoran MJ, Funte LR, Kater SB, Haydon PG (1993) Contact with a synaptic target causes an elevation in a presynaptic neuron's resting calcium set-point. Dev Biol 158:163–171

Index

Achatina 60, 72, 96–97, 119
Acheta 32, 137
Acklin and Nicholls 1990 120
adhesion 67, 99, 122, 145–146, 160
age and regenerative potential 9, 20, 24, 59
Aguayo et al. 1991 7
Allen et al. 1981 78
Allison and Benjamin 1985 132
Allison and Benjamin 1985 133
Allison and Benjamin 1985 72
Aloe et al. 1994 71
Altrup 1982 72
Alvarez-Buylla and Nottebohm 1988 1
Ambron et al. 1989 105
Ambron et al. 1989 105
amphibian 6, 88, 89, 144
Amphioxus 39
Anderson and Schwab 1982 14
Anderson and Spencer 1989 14
Anderson et al. 1980 29, 88
Aplysia 35, 38, 41, 47–48, 49–50, 53, 54, 56, 58, 59, 61, 76, 78–79, 88, 100, 104–105, 116, 117, 121, 127, 131, 134, 143, 145–148, 154, 160
apoptosis 23, 53, 119
Arbas and Weidner 1991 33
Aréchiga et al. 1986 127
asexual reproduction and regeneration 18
Asterias 39
Atashi et al. 1992 99
Atwood and Wojtowicz 1986 161
Atwood et al. 1989a 161
Atwood et al. 1989a 62
Atwood et al. 1989b 61
Atwood et al. 1989b 71
auditory systen (insect) 139
autotomy, 9, 13, 32–33, 39
Avery and Horvitz 1987 23
avoidance, in neurite interactions 160
axon distal segment survival
- annelid 63–65
- crustacean 30–32, 62
- gastropod 66
- innervation of muscle 31–32
- role of glia 55
axon fusion 27, 28, 29, 31, 41, 62–66
axon membrane repair 44–51
axonogenesis 32, 105
axotomy
- changes in electrophysiological properties 51–52
- cytoskeleton 51–52
- early responses 42
- gastropod axon responses 48
- initiation of sprouting 72
- soma response 52; 55–58
Azhderian et al. 1994 56, 59

Bacon and Blagburn 1992 137, 138
Bacon and Murphey 1984 135
Baguna 1976 18
Baguna and Ballister 1978 16
Baguna et al. 1988 18
Baguna et al. 1989a 18
Baguna et al. 1989b 18, 19, 20
Baier and Bonhoeffer 1991
Baier and Bonhoffer 1992 121, 123
Baier and Bonhoeffer 1994 111
Bailey and Chen 1989 135
Bailey.et al. 1992 135
Baker and Croll 1991 55
Baker et al. 1993 55, 72, 96, 97
Bamburg et al. 1986 91
Bandtlow et al. 1993 94
Bannatyne et al. 1989 51, 72
Baptista and Macagno 1988 127
Baptista et al. 1990 6
Barker et al. 1982 115
Bartel et al. 1989 42
Barthalay et al. 1990 104
Bastiani et al. 1987 100
Bate 1976 111
Bateson 1894 33
Baylor and Nicholls 1971 29, 73
Beadle et al. 1988 111
Beck et al. 1990 121
Bedlack et al. 1992 44
behavior

- failure to recover 134–135
- flatworms 20
- hypersensitivity 131–132, 134
- leech, in absence of regeneration 29, 129
- oligochaete 26–28
- sensitization 131
- tentacle reflexes 38
Belardetti et al. 1987 82, 84
Benbassat and Spira 1993 52, 61, 66
Benbassat and Spira 1994 61, 66
Benjamin and Allison 1985 72, 132
Benjamin and Allison 1987 72
Bentley and Toroian-Raymond 1986 80
Bentley et al. 1991 82, 88
Berdan and Bulloch 1990 157
Berdan and Easaw 1992 54
Berdan and Ridgway 1992 115
Berdan et al. 1989 70
Berdan et al. 1990 56, 57, 70
Berdan et al. 1993 42, 44, 48
Bickell-Page and Mackie 1991 13
Birse and Bittner 1976 27
Bittner 1977 32
Bittner 1988 61, 62, 71
Bittner 1991 62
Bittner and Baxter 1991 161
Bittner and Johnson 1974 31
Bittner et al. 1974 29
bivalves 34
Bixby 1989 121
Blackshaw et al. 1982 3, 73
Blair et al. 1987 80
Blanco 1990 67, 69
Blanco and Lane 1990 67
blastema 18, 71
Bliss 1960 9, 32
blueprint hypothesis of pathfinding 144
Blundon et al. 1990 11, 62
Bode 1992 15
Bode et al. 1988 15
Bode et al. 1990 16
Bodenstein 1957
Bodmer et al. 1984 145
body part regeneration 9, 24, 25, 27
- arthropods 32
- cephalopod 34
- cnidaria 14, 16
- ctenophora 16
- echinoderm 39
- gastropod 35–36
- nemertea 21
- oligochaetes 25
- platyhelminthes 18
- polychaetes 24
Bollner 1989 39, 71
Bollner et al. 1992 39
Bollner et al. 1993 39, 71
Bollner et al. 1995 40
Bombyx 117
Bouton and Bittner 1981 31, 62
Bovolenta et al. 1991 4
Bowling et al. 1978 73
Bradacs et al. 1991 161
branching pattern of neurites 120
Brandstätter et al. 1992 72
Braun and Stent 1989 4
Bray 1991 112
BRDU (bromodeoxyuridine) 58, 70
Breidbach et al. 1992 106
Briedbach 1988 32
Broadie and Bate 1993a 157, 158
Broadie and Bate 1993b 157
Broadie and Bate 1993c 157
Brondsted 1969 18
Brown et al. 1981 3
Brusca and Brusca 1990 39
budding, fragmentation and reproduction 9, 18
Budnick et al. 1989 72
Bulloch 1984 72
Bulloch 1985 10, 116, 132
Bulloch 1987 116
Bulloch 1989 72
Bulloch and Hauser 1990 115, 116
Bulloch and Kater 1981 51, 132
Bulloch and Kater 1982 132, 134
Bulloch and Ridgway 1989 3, 10
Bulloch et al. 1980 38
Bulloch et al. 1984 127
Bullock and Horridge 1965 34
Büschges and Pearson 1991 139
Büschges et al. 1992a, b 139–141
Byers 1970 56

c-fos 58
Ca^{2+} *see* Table of Contents
- and cell adhesion molecules 99
- and substrate interactions 104
- changes after axotomy 42–43
- changes in conductances 160
- effect on cytoskeleton 44
- setpoint hypothesis 157
Caenorhabditis elegans 23, 110–111, 112, 153
Cagan and Ready 1989 88
Cagan and Zipursky 1992 88
Camhi and Macagno 1991 63, 64
Campanelli et al. 1992 157
Cancalon 1985 62
Candia-Carnevali et al. 1989 39
Carbonetto and Muller 1977 63
Carbonetto and Muller 1982 62

Carcinus 140
Cardisoma 112
Caroni and Schwab, 1988 4
Carroll and Kempf 1994 35
Carrow and Levitan 1989 145, 146
Casagrand and Ritzman 1991 137
Cash and Carew 1989 35, 143
Cash et al. 1992 152
Catarsi and Drapeau 1992 158
Caudy and Bentley 1986 111
Cayre et al. 1994 32
cell adhesion molecule (CAM) 7, 146
cephalopod 34, 56
cerci 135
Cervero and Laird 1991 42
Chalfie et al. 1985 23
Chamak and Prochiantz 1989 121
Chang et al. 1992 99
Chase 1986 35
Chase and Kamil 1983 35
Chase and Rieling 1986 35
Chase and Tolloczko 1987 35
Chen and Levi-Montalcini 1970 119
Chiasson et al. 1994 94, 97
Chiba and Murphey 1991 51, 126, 135, 157
Chiba et al. 1988 135
Chiba et al. 1993 152
Chien et al. 1993 80
Ching et al. 1993 158, 159
Chiquet and Acklin 1986 81
Chiquet and Nicholls 1987 81, 101
Chiquet et al. 1988 67, 103, 103, 106
Chitnis and Kuwada 1991 80
Ciona intestinalis 39
Clark et al. 1962 24, 26
Clement et al. 1983 148
Cnidaria (coelenterates) 4, 15
cockroach 44, 46, 67, 69, 83, 85, 99, 105, 109, 119, 135, 148–151
Cohan 1990 90, 91
Cohan and Kater 1986 90, 91
Cohan et al. 1985 87
Cohan et al. 1987 82, 94, 134
Cohan et al. 1990 155
Cohen and Baker 1988 135
Cohen and Jacklet 1965 56
Cole et al. 1989 29
collagen 100, 104
collapse of growth cones 104, 115
collapsin 99
Collett 1994 89
competition
- between neuronal processes 150
- responses to deafferentation 135
Con A 100, 103, 104, 119, 145–146
Condic and Bentley 1989a-c 80, 98
conditioned medium 104, 111–115
conditioning lesion 59
connectin 99
Conner and Smith 1994 56
Constantine-Paton et al. 1990 89
crab 9, 33, 111
crayfish 32, 33, 42–43, 52, 56, 62, 71, 127, 137, 148
Crepidula 35, 37
cricket 32, 52, 56, 135–138, 139, 153
critical period 33
Croll and Baker 1990 51
Croll and Chiasson 1989 35
Cryptomphalus 70–71
ctenophore 15, 16
culture conditions 99, 111–123
Curran and Morgan 1995 58
cyclic AMP 56, 59
cytochalasin 78, 80
cytoskeleton
- adhesion 99–100
- and calcium regulation 91
- and calmodulin, 94
- axon repair 72–73
- changes after axotomy 44
- neurotransmitter responses 82
- solubilization 78
- stabilization/destabilization 78–79
- transmembrane linkages 121
Czeh et al. 1978 55

Dai and Peng 1993 157
Dale et al. 1987 154
Daphnia 88
Dash et al. 1991 89
Davenport and Kater 1992 92
David and Gierer 1974 15
David and Hager 1994 16
Davies et al. 1985 20
Davis 1989 88
Davis and Murphey 1994 158
Davis et al. 1992 54, 55
deafferentation 131–140, 135
defasciculation 107, 108, 110
degeneration
- distal axon segment 30–31
- muscles following autotomy 33
- prevention, by axon survival 62
- Wallerian 61
Delaney and Zucker 1990 82
Denberg and Caldwell 1992 149
Denburg 1982 150
Denburg 1985 127, 151, 152
Denburg 1988 150
Denburg 1989 105

Denburg and Caldwell 1992 32, 150, 151
Denburg and Norbeck 1989 105, 150
Denburg and Whitington 1978 2, 33
Denburg et al. 1977 150
Denburg et al. 1983a,b 104, 150
Denburg et al. 1987 104, 150
Denburg et al. 1988 151
Denburg et al. 1989 105
depolarization; following injury 41–42
DeRiemer et al. 1983 63
Desarmenien and Spitzer 1991 89
development
- and regeneration, compared 7, 125–127
- *Aplysia* 143
- cockroach 105, 111
- critical period 125–126
- *Drosophila* 99, 151
- insect 119
- insect neuromuscular innervation 148
- leech 108, 128
- locust 104, 117
- reexpression of cues 106, 123, 150
- snail 128
differentiation, extrinsic/intrinsic factors 125
distal axon segment
- crustacean, 31
- electrical connection with proximal segment 62, 66
- release of transmitter 61–62
- sprouting 54, 62
- survival 61–62
Doherty et al. 1991 100
Dotti et al. 1988 120
Drapeau 1990 158
Drapeau and Sanchez-Armass 1988 158
Drapeau and Sanchez-Armass 1989 158
Drapeau et al. 1989 158
Drewes and Fourtner 1990 25, 27
Drewes et al. 1988 27, 64
Drosophila 32, 72, 78, 88, 89, 100, 103, 104, 105, 112, 116, 119, 125, 151–152, 153, 154, 157
Dugesia 18, 19, 20

earthworm 9, 26–28, 44, 48, 64–66, 89
Ebberink et al. 1989 116
ecdysone 117
ectopic transplants, arthropod 29
Edwards 1988 33
Edwards and Palka 1971 33, 137
Edwards and Palka 1974 135
Edwards et al. 1989 33
Egid and Lnenicka 1993 91
Egid and Lnenicka 1994 91
Eichler et al. 1972 34
Eisen et al. 1986 92
Eisenia 26, 64
electric fields 44
electrical synapses 88, 132
Elliot and Muller 1982 72
Elliot and Muller 1983 72
Elphick et al. 1989 39
Elvin and Koopovitz 1994 18
Ely and Vélez 1982 32, 148
Emson and Wilkie 1980 32, 39
endoreplication 35
epidermal growth factor 112
epimorphic processes 18
ethidium bromide 67
extracellular matrix 67, 69, 99, 100, 105, 119
eye regeneration 70

Faisst et al. 1980 20
Falls et al. 1993 157
fasciclins (semaphorins) 99, 103, 110
fasciculation 4, 106, 107, 108, 110, 144
Fawcett 1992 123
Fazeli 1992 53
Fekete and Brockes 1987 2, 150
Fenger et al. 1994 15
Fernandez and Flores 1978 37
Fernandez-de-Miguel et al. 1992 155
Fessler and Fessler 1989 104, 105
Fessler et al. 1987 103
fibronectin 100, 104, 119
Fields et al. 1990 92
filopodia 151
fish 6, 59, 62
Fishman et al. 1990 48, 49
fission 9
flatworm 1, 16–21
flight (locust) 139
Flores and de Iraldi 1973 35
Flores et al. 1992 35, 37, 38, 70, 71
Forscher and Smith 1988 78
Forscher et al. 1987 78, 79
Frank et al. 1975 27, 28
Franquinet 1979 19
Fredman and Nutz 1988 59
French and Kristan 1992 97
French and Kristan 1994 29
French et al. 1992 98
Friedlander and Levinthal 1982 32
Fromherz and Schaden 1994 79
Fromherz et al. 1991 122
Fuchs et al. 1981 127
Fuchs et al. 1982 153
Fuentes et al. 1994 15, 16

Fujisawa 1989 15
Funte and Haydon 1993 156, 157

Gabel et al. 1989 104
Gallant 1988 48
Gallant 1992 52
ganglion excision 141
ganglion regeneration
- gastropod 35, 38
- polychaete 24
- tunicate 39–40
ganglion transplantation 24, 26
- insect 29, 33
- platyhelminthes 20
- polychaete 24
- oligochaete 26–27
- snail 39
Gao and Macagno 1987a,b 160
Gao and Macagno 1988 29, 128
gap junctions 87–88
GAP-43 120
- reexpression in regeneration 55
- and axon repair 58
Garcia-Arraras et al. 1994 39
Garcia-Fernàndez et al. 1991 20
gene activation 41–42, 58
genes, and neuromuscular innervation 153
Geraerts and Joosse 1989 116
Gertler et al. 1989 78
Ghirardelli 1965 19
giant axon (giant fiber)
- *Aplysia* 47–50
- annelid 27–28
- arthropod 29
- cnidaria, 13
- cockroach 44, 49, 99
- earthworm 27–28, 44, 49–50
- insect 135
- squid, and axon seal 44, 48, 49
Gillette 1991 35
Gilly and Brismar 1989 52
Glanzman et al. 1989 41, 131
Glanzman et al. 1990 154
Glanzman et al. 1991 146, 160
Glover and Mason 1986 97
glycoproteins 100, 104, 105, 116
Gnegy 1993 94
Go and Hotta 1992 32, 105
Goedert et al. 1991 91
Goldberg and Burmeister 1989 75, 79, 80
Goldberg and Burmeister 1992 76
Goldberg and Kater 1989 72
Goldberg and Schacher 1987 121, 151, 160
Goldberg et al. 1988 35
Goldberg et al. 1990 97, 128
Goldberg et al. 1991 35, 78, 97, 117, 121
Goldberg et al. 1992 35, 97
Golding 1967a-d 24
Golding 1974b 24
Golding 1987 24
Golding 1974a 21
Gomot et al. 1992 116
Gonzalez and Silver 1994 2
Goodman 1994 98
Goodman and Heitler 1979 51, 52, 55
Goodman and Shatz 1993 4, 160
Goransson et al. 1988 32, 148
Gordon-Weeks 1991 76, 120
Gordon-Weeks and Mansfield 1990 91, 120
Goslin et al. 1988 120
Goss 1969 6, 9, 24, 34
Gottman et al. 1991 82
Govind 1992 29, 161
Govind et al. 1988 33
Govind et al. 1992 31, 54, 71
gradients of pathfinding cues 111, 121–123
Graf and Cooke 1990 78, 111, 112, 116
Graf and Cooke 1994 92
Gras 1984 27
grasshopper 79, 80, 88, 97, 100, 108, 110, 135
Grau and Cooke 1992 78
Graziadei and Monti-Graziadei 1978 6
Graziadei and Monti-Graziadei 1980 2
Greger and Berking 1991 16
Grenningloh et al. 1990 100
Grigoriev and Spencer 1994 13
Grimm-Jorgensen 1983 116
Grimm-Jorgensen 1987 116
Grinvald and Farber 1981 82
Grossfeld and Hansen 1987 71
Grossfeld et al. 1994 56, 61
growth cone
- anatomy 75–78
- antibody interference 108
- behavior 73–82, 97
- cytoskeleton 78–80
- differentiation 97–98
- environmental cues 98
- growth state 86–87
- ionic channels, 80, 82–94
- navigation 80–81
- responses to neurotransmitters 94
- substrate effects 81, 84–86
- veils 79
growth factors 9, 113–117
- insulin-like 59
- nerve growth factor 59–60
Grumbacher-Reinert 1989 81, 119, 121
Grumbacher-Reinert and Nicholls 1992 85, 86, 91, 92

Gu and Muller 1990 73, 161
guidepost cells 99
Gustaffson and Eriksson 1992 18
Gustafson and Toneby 1970 39
Gustafsson 1992 18
Gustafsson and Eriksson 1992 18
Gustafsson and Reuter 1990 18
Guthrie 1967 31
Guthrie 1985 15
Gutmann et al. 1941 54

Hadley et al. 1982 51
Hadley et al. 1983 132
Hall and Lloyd 1991 116
Hall and Russell 1991 23
Hall and Sanes 1993 158
Hall et al. 1989 52
Halpern et al. 1991 99, 103, 146
Hamamoto et al. 1988 123
Hamelin et al. 1993 110
Haralson and Groff 1975
Harrelson and Goodman 1988 100
Harzsch and Dawirs 1994 29
Hawver and Schacher 1993 146
Hay 1966 21, 34
Haydon 1988 155
Haydon and Kater 1988 125, 148
Haydon and Zoran 1989 148
Haydon and Zoran 1994 125, 126, 156
Haydon et al. 1984 72
Haydon et al. 1985 120
Haydon et al. 1987 94
Heamopsis 28
Hedgecock et al. 1987 110
Hedgecock et al. 1990 23, 110, 111
Hekimi and Kershaw 1993 110
Helisoma 38, 42, 44 48, 54, 56–57, 59, 70, 72, 86, 91–95, 97, 98–99, 101–102, 104, 113, 115–116, 117, 125, 128, 132–134, 155, 156–157
Helix 35, 37, 83–84, 114–115
Helobdella 28
hemocytes 1
Henderson 1983 158
Herlant-Meewis 1964 24, 26, 27
Hernandez-Nicaise 1973 16
Herrup and Busser 1994 58
heterophilic binding 99
Hickmott and Carew 1991 35, 143
Hildebrand 1985 33
Hinrichsen 1993 94
hirudinea, *see* leech
Hirudo 28
Hitchcock and Raymond 1992 6, 123
Holder and Clark 1988 108
Holliday and Spitzer 1990 88, 89
homeodomain protein 153
homeotic regenerates 33, mutants 33
homophilic binding 99
Hopkins et al. 1979 32
Hortsch and Goodman 1991 100
Howes et al. 1989 67
Howes et al. 1991 69, 82, 119, 120
Hoy et al. 1967 31, 61, 63
Huber 1987 139
Huber and Thorson 1985 139
Huet 1975 39
Hulsebosch and Bittner 1980 9, 27
Hume et al. 1983 82
humoral (blood-borne) factors and regeneration 24
Hunt and Vélez 1982 32, 148
Hunter et al. 1991 121
Hydra 13, 14, 15, 16, 19
- head activator 14, 16, 19
- head inhibitor 16
Hydractinia 15
hyperinnervation
- crustacean muscle 32
- regenerating insect muscle 32

immune response 71
immunoglobulin-like molecules 99
implanted neurons 99
induction
- effects of astrocytes, 7
- planarian 19–20
inhibitors of growth and differentiation 119
injury discharge 41–42
input carrying capacity 160
insulin-like growth factor 116–117
integrins 100
invertebrate model systems, rationale 10
ion channels
- in development 88–89
- in spontaneous activity 90–91
- regulation in growth cone 82–87
Ito and Hotta 1992 32
Ivins et al. 1991 92

Jacklet and Cohen 1967 31
Jacob 1984 35, 143
Jacob and McQuarrie 1991 58
Jacob and McQuarrie 1993 59
Jansen and Nicholls 1972 29, 73, 129, 134, 158
Jansen et al. 1974 131, 134
Jay and Keshishian 1990 110
Jellies et al. 1987 4, 97

Jellies et al. 1994 108
jellyfish 58
Jeppesen 1976 37
Jessell 1988 99
Jessell et al. 1990 104
Johansen et al. 1985 28
Johansen et al. 1992 108
Jones and Bulloch 1988 54
Jones et al. 1987 72

Kaczmerek et al. 1979 116
Kalmring 1975 139
Kämper and Murphey 1987 150, 153
Karlstrom et al. 1993 100
Kater 1985 98, 99
Kater and Mattson 1988 80
Kater and Mills 1991 75, 82, 92, 95, 157
Keller and Schacher 1990 146
Kempf et al. 1992 35
Kennedy 1971 137
Kennedy and Bittner 1974 54
Keshishian 1993 32, 125
Keshishian et al. 1994 111, 125, 151, 152
Killian et al. 1993 136, 137
Kim and Wu 1991 119, 120
Kim et al. 1991 80
Kimble 1981 23
Kimmel and Carlyon 1990 20
Kiortsis and Moraitou 1965 24
Kiortsis and Trampusch 1965 34
Kirk and Govind 1992 88, 148
Kits et al. 1990 116
Kleinfeld et al. 1988 121
Kleinfeld et al. 1990 127
Klose and Bentley 1989 4, 80
Knobloch and Steel 1988 32
Koizumi and Bode 1991 15
Koizumi et al. 1989 14
Koizumi et al. 1990 14, 15, 18
Kolodkin et al. 1992 100
Koopowitz 1989 16
Koopowitz and Holman 1988 20, 62, 63
Koopowitz et al. 1975 20, 63
Koopowitz et al. 1976 20, 63
Koyama and Kusunoki 1993 39
Krasne 1987 137
Krause and Bittner 1990 11, 66
Krause et al. 1991 66
Krause et al. 1994 44, 48, 49, 50
Krik and Bulloch 1992 60
Krishnan et al. 1993 153
Kristan et al. 1993 158
Kruk and Bulloch 1992 59
Krull et al. 1994 103
Kuffler and Muller 1974 144
Kuno and Llinas 1970 52
Kupfermann 1974 38
Kuwada 1981 52
Kuwada and Wine 1981 52

LaBerge and Chase 1992 35
lachesin 99
Lakes 1990 139
Lakes et al. 1990 134, 139
Lakes-Harlan and Pollack 1993 99
laminin 67, 100–104, 115, 119, 121
lamprey 44–45, 135
Lander 1989 99
Landmesser et al. 1990 100
Lange 1920 34
Lankford and Letourneau 1989 91
Lasek et al. 1974 61
Lauder 1993 39, 94, 95
Le Gall 1983 35, 37
Le Gall et al. 1983 35, 37
Leah et al. 1991 42
learning 53, 56, 58
lectins 104–106, 150
leech 6, 28–29, 51, 62–63, 67–68, 72–73, 75, 79, 81, 85–86, 88, 97, 101, 106–109, 119, 121, 127, 128, 131, 134, 144–145, 155, 158–161
Lees and Beadle 1991 89, 119
Lender 1956 19
Lender and Bouchard-Madrelle 1964 39
Letourneau 1976 33
Letourneau et al. 1991 75
Lichtman 1982 3
Lichtman and Balice-Gordon 1989 160
Lieberman 1971 56
Lin and Levitan 1987 146
Lin et al. 1989 91
Linckia 39
Lnenicka and Atwood 1989 161
Lnenicka and Murphey 1989 135
Lnenicka et al. 1991 161
lobster 71, 90, 111, 161
locomotion pattern generator 139–140, 141–143, 153
locust 51–52, 139
Locusta 139
Loer and Kristan 1989 97
Loer and Kristan 1989 97
Loer et al. 1987 28, 97
Lohof et al. 1992 78
LoPresti et al. 1973 80, 88
LoPresti et al. 1974 88
Lubet 1971 35
Lumbriculus 27
Lumbricus 27, 64–65
Lyckman and Bittner 1992 65

Lyckman et al. 1992 62, 65
Lymnaea 37, 52, 72, 113, 116–117, 128–130, 133
Lyngdoh and Tandon 1992 18

Macagno et al. 1985 131, 144
Macagno et al. 1986 97
Macagno et al. 1989 29, 144, 160
Mackie 1980 13
Maetzold and Bulloch 1986 72
Mahanthappa 1994 6
mammal 2, 10, 11, 75, 112, 144
Manduca 103
MAPS (microtubule associated protein) 78, 91, 120
Marcum and Campbell 1978 15
Marom and Dagan 1987 82, 83, 84
Martelly and Franquinet 1984 18, 19
Mason and Muller 1983 28
Masuda-Nakagawa and Nicholls 1991 29, 67, 103
Masuda-Nakagawa and Wiedemann 1992 67, 81, 101
Masuda-Nakagawa et al. 1988 101, 119, 121
Masuda-Nakagawa et al. 1990 67, 68, 108
Mattson and Kater 1987 92
Mattson and Kater 1988 104, 105
Mattson et al. 1988a 95
Mattson et al. 1988b 92, 93, 104, 120
Matzner and Devor 1994 51, 52
Maule et al. 1993 18
Mauthner neurons 44
May 1933 34
May et al. 1987 35, 143
Mayford et al. 1992 100
Mayford et al. 1992 146
McAllister et al. 1983 143
McCarthy et al. 1994 58
McClellan 1992 135
McCobb et al. 1988a, b 94, 97
McFaddan and Spencer 1992 15
McFarlane et al. 1989 14
McGlade-McCulloch and Muller 1989 67
McGlade-McCulloch et al. 1989 67
McGlade-McCulloh and Muller 1989 29
McKerracher and Hirscheimer 1992 58
McLane et al. 1994 89
McLean and Edwards 1976 33
McQuarrie and Grafstein 1982 59
McQuarrie et al. 1977 59
McRorie and Zipser 1988 108
McTeague and Govind 1993 33
McVean 1975 32
Meadows et al. 1994 100
Mearow et al. 1994 115
Meinertzhagen 1993 89
Meiri and Burdick 1991 83
Meiri and Gordon-Weeks 1990 120
Meiri and Grafstein 1984 44
Meiri et al. 1981 44, 51, 83
Melampus 37, 38, 70, 131–132, 141–143
Merz and Drapeau 1992 158
Merz and Drapeau 1994 29
metamorphosis 32
Microstomum 20
Millar and Ratcliffe 1994 71
Miller and Hadley 1991 67, 102, 104, 108, 115
Miller et al. 1992 153
Mills and Kater 1990 54
Mimura 1993 89, 125
Mittenthal 1980 33
Moffett 1975 32
Moffett 1987 33
Moffett 1990 141
Moffett 1991a 37
Moffett 1991b 38
Moffett 1992 38
Moffett 1993 37
Moffett 1995 34
Moffett and Austin 1982 38, 143
Moffett and Bale 1993 70
Moffett and Ridgway 1988a 38, 141
Moffett and Ridgway 1988b 38, 143
Moffett and Snyder 1985 38, 131, 141, 143
Moffett et al. 1995 60, 119
molluscs 34–38
molting, coordination of appendage regeneration 32
Moment 1951 24
Montell and Goodman 1988 103
Moody et al. 1991 88
Morgese et al. 1983 67
morphallaxis 18, 21, 27
Muller 1988 28
Müller 1991 16
Muller and Carbonetto 1979 29, 63
Muller and Gu 1991 29, 71, 72, 161
Muller and Scott 1979
Muller and Scott 1980 72
Muller et al. 1981 28
Muller et al. 1987 29, 73, 128
Muramoto 1993 42, 43
Murphey 1986 3, 10
Murphey et al. 1976 137
Murphey et al. 1980 135
Murphey et al. 1983a,b 150
Murphy and Kater 1980 51
Murphy et al. 1984 132
Murrain et al. 1990 94
muscle reinnervation

- arthropod 30–32
- glycoprotein matching 104–105
- homophilic binding 100
- insect 103
Muskavitch and Hoffman 1990 112
mutants 23, 110, 112, 137, 153
Myers et al. 1990 132, 52

N-cadherin 99
Nagasawa et al. 1986 117
Nardi and Vernon 1990 2
Needham 1952 34
Needham 1965 32
nematoda (roundworms) 4, 23
nemertia (ribbon worms, rhynchocoels) 21
Nereis 24
nerve crush 67, 68, 100
netrins 111
network reconstruction 128–129
neural cell adhesion molecule (N-CAM) 99
neural development
- arthropod 29
- cnidaria 15
- gastropod 35
- leech 28
- nematoda 23
- platyhelminthes 18
neural induction 6
- crayfish 33
- echinoderm 39
- gastropod 37
- nemertea 21–22
- platyhelminthes 19
- polychaete 24
neuroblasts 4
neurogenesis 33
- amphibians, forebrain, retina 6
- birds 6
- *Caenorhabditis* 23
- continuous, indeterminate lifespan 9
- cyclical,song control nuclei 4
- fish, forebrain 6
- gastropod 143
- leech 29, 108
- mammals, olfactory 6
- trophic controls 6
neuromodulators, release following injury 59–60, 73
neuromuscular innervation, and lectins 104
neuron crush 132
neuron replacement 4
- gastropod 35
- *Hydra* 15
- insect 33
- platyhelminthes 18
- tunicate 39
neuroparsin 117
neurotactin 104
neurotransmitters in regeneration 94–97
- chordata 39
- cnidaria 14–15
- echinodermata 39
- gastropods 94–95
- platyhelminthes 18–20
- serotonin, gene activation 41
Newland et al. 1993 120
NGF 113, 116
Nicholls 1986 28, 88
Nicholls and Hernandez 1989 82, 89
Nicholls and Van Essen 1974 28
nonneural cell responses
- amebocytes 70
- explanted cockroach ganglia 67
- glia 70
- hemocytes 76–69, 70–71
- incorporation of glial nuclei 71
- microglia in leech 67
- neuron growth 119
- phagocytic cells 70
- sheath cells 70, 119
Norbeck et al. 1992 111, 121
Nose et al. 1992 100
Notoplana 20–21, 63
Nottebohm 1981 6
Nottebohm and Nottebohm 1978 6
novel synaptic connections 128–129

O'Connor et al. 1990 75, 76, 77, 79
Okabe and Hirokawa 1991 75
Okamura and Takahashi 1993 88
Okamura et al. 1993 39, 88
olfactory system 2, 35–37
oligochaetes 25–28
Olive 1974 24
Ophryotrocha 24

Page 1992a, b 35
Page 1993 35
Palka 1993 7, 97
Palka and Edwards 1974 137
Palka et al. 1992 4, 99
Palladini et al. 1988 18
Pallas and Hoy 1986 127, 134, 139
Palmberg 1986 18, 20
Palmberg and Reuter 1990 20
parabiosis 21
Parente and Lentz 1986 66

Parnas et al. 1984 161
Parnas et al. 1991 55, 62
Parsons et al. 1991 122
Passani et al. 1991 108, 137
Patel et al. 1987 100
pathfinding 1, 95–97, 106–111
Patterson 1988 99
Pearson and Bradley 1972 148
Peinado et al. 1987a 109
Peinado et al. 1987a, b 108
Pellegrino et al. 1984 52
Periplaneta 137
Perry et al. 1993 3, 52
Perry et al. 1993 52
Pestarino 1993 39
Pestarino et al. 1993 39
Pfannenstiel 1984 24
Pfenninger et al. 1990 78
photolithography 121
Physella 116
Pilkington and Benjamin 1986 52
Pinter et al. 1991 55
pioneer neuron 4, 144
- characterization of cues 108–111
- grasshopper 80, 98
- responses to gradients 111
Piovant and Lena 1988 116
Pitman and Rand 1982 72
Pitman et al. 1972 52, 55
plasticity
- *Aplysia*, N-CAM 100
- of central connections 139
- of neurons 126, 128
- of the nervous system 3, 10, 162
Platyhelminthes (flatworms) 16, 18, 19
Plickert 1989 16
Polak et al. 1991 94
polychaetes 23–25
polyethylene glycol 122
Polyorchis 15
Price 1977 141
Price 1977 38
programmed cell death *see* apoptosis
proteoglycans *see* glycoproteins
prothoracicotropic hormone 117
Purves et al. 1988 120

Rankin and Carew 1987 35
Raper et al. 1983 92
reactive synaptogenesis 3
Reber-Muller et al. 1994 58
recapitulation of development 1, 2, 4
reexpression of cues 7, 123, 150
refinement of connections 150–152
Rehder and Kater 1992 92
Reichardt and Tomaselli 1991 121
Reichert 1994 153
Reinecke et al. 1993 59
Reis and Ross 1973 58
remodeling of connections 125–134
- central connections 6, 135–143
- peripheral nerves 6, 148–152
retraction bulbs 51
retrograde transport 55–56, 59
Reuter and Kuusisto 1992 18
Reuter and Palmberg 1989 18
Reynolds and Weiss 1992 6, 112
ribbon worms 1, 21–23
Richards et al. 1983 122
Richmond et al. 1991 155
Ridgway and Bulloch 1991 116
Ridgway et al. 1989 116
Ridgway et al. 1991 60, 113, 115, 116
Roberts and Mackie 1980 13
Roederer and Cohen 1983 52, 56
Roger et al. 1992 128, 129
Ross et al. 1994 66
Ross et al. 1975 58
Ross et al. 1987 82, 85, 122
Ross et al. 1988 84, 85, 86, 90
Ross et al. 1994 61
Rossiter et al. 1994 58
rotifers 4
Rotshenker 1979 3
Roubos et al. 1988 35
Rubel et al. 1995 6
Rutishauser et al. 1988 99
Ryerse 1989 88

Sabry et al. 1991 75, 76, 78
Salo and Baguna 1986 18, 19
Sanes 1989 99
Sarsia 14
Sawyer 1981 28
Schacher et al. 1979 88
Schachner 1991 100
Schaller et al. 1989 15, 16
Schaller et al. 1990 15
Schildberger et al. 1986 139
Schistocerca 88, 139
Schmiege et al. 1992 33
Schmitz 1989 139
Schmitz et al. 1988 139
Schuch et al. 1989 100
Schultze 1899 39
Schwartz 1987 58, 119
Scott and Kirk 1992 38
Seaver et al. 1991 108
selection pressure and reproduction 8
semaphorins *see* fasciclins

sensory neuron regeneration
- arthropod 33
- insect 105–106
- leech 106–109
- vertebrate, and NGF 115
serotonin *see* neurotransmitters
Seshan and Bittner 1987 29
Seshan and Levi-Montalcini 1973 119
sheath cells 67, 68–69, 119
Sheller et al. 1991 61
Sheng and Greenberg 1991 42
Sheng et al. 1991 89
Shepherd and Murphey 1986 137
Shepherd et al. 1988 135
sibling neurite bias hypothesis 151
Sidelnikov and Stepanov 1991 60, 119
Sidelnikov and Stepanov 1992 60, 119
Sigurdson and Morris 1989 84
Silver et al. 1990 82
Singer et al. 1979 144
Singla 1978 13
Sink and Whitington 1991a, b 152
Skene 1992 55
Skinner 1985 32
Skinner and Graham 1972 32
Smalheiser and Crain 1984 151
Smith and Howes 1987 61, 67
Smith et al. 1990 69
Smith et al. 1991 69, 71
Snider and Johnson 1989 9
Snyder and Moffett 1987 131
Snyder and Moffett 1990 131, 141
Snyder and Moffett 1990 143
Sonetti and Bianchi 1993 59
Sonetti et al. 1992 117
Spencer and Satterlie 1981 13
Sperry et al. 1973 19
Spira et al. 1987 85, 99, 120, 121
Spira et al. 1993 47–50, 54
Spitzer 1991 82, 83, 88, 155
sprouting 92–97
- calcium ionophore A23187 92
- calcium second messengers 94
- calmodulin 94
- culture factors 116–119
- cut axons 53
- growth cone collapse 92, 94
- inhibition 117
- regulation by substrate and Ca^{2+}
- uninjured neurons 72–73
squid 44, 48–49, 52
Stasek 1967 32
Stefano 1992 71
Steinhardt et al. 1994 44
Stent et al. 1992 28, 97
Stewart and Atwood 1992 127
Stewart and McCaig 1994 89
stop growth cues 4, 9, 128
Strandburg and Krasne 1985 137
Strautman et al. 1990 44, 45
Strittmatter and Fishman 1991 99
Strongylocentrotus 112
Stuart et al. 1987 28
Stuermer et al. 1992 2, 7
substrate
- attachment 119
- components 100–104
- growth cone ion channels 84–86
- growth cone morphology 81
- growth cone collapse 97
- sprouting 119–121
- supply of trophic factors 111
Syed et al. 1990 128, 130
Syed et al. 1992a 99, 128
Syed et al. 1992b 153
Syed et al. 1993 129
synapse
- Ca^{2+} channel distribution 156
- electrical activity 157
- elimination 127
- excitation-contraction coupling 155
- postsynaptic responses 158–159
- prospective 125–127, 144
- protein synthesis 155
- retrospective 125–127, 144
- strength 135, 137
synaptogenesis, extracellular matrix antigens 32

Taghert et al. 1982 92
target
- trophic effects 51, 55–56, 58, 61062, 72–73
- recognition 62–63, 75, 89, 98, 101–105, 108, 115, 119, 123, 137, 140–144
tau 120
tenascin 100, 103
Tenebrio molitor 106
tentacle regeneration 70
Thoenen 1991 116
Thomson 1990 120, 125
Thorey and Zipser 1991 28, 29
thrombospondins 100
Titmus and Faber 1990 51
Tiwari and Woodruff 1992 115
Todorov et al. 1992 44, 66, 89
Tolbert and Oland 1990 103
Tomlinson and Ready 1987 88
Tornqvist et al. 1994 55
Torrence et al. 1989 28
Tosney and Landmesser 1985 92

transplantation
- of brains 24, 26
- of limbs 152
- of neurons 128, 153
Travis 1992 6
Travis 1994 98
Treherne et al. 1987 69
Treherne et al. 1988 69
Tribut et al. 1991 52
trophic factors 111, 119
Truman 1990 32, 105
Tublitz 1993 127
tubulin, in axon repair 58
Tucker 1959 21
tunicates 4, 39–40, 71
Turrigiano et al. 1994 90, 140

Uemura et al. 1994 39
Ushkaryov et al. 1992 125

Van Berkum and Goodman 1995 94
Van Essen and Jansen 1977 144
Van Vactor et al. 1993 153
Vanhems and Delbos 1989 104
Vanhems et al. 1990 117
Vanhems et al. 1993 119
Venstrom and Reichardt 1993 100, 101, 121
ventilatory control pattern generator 140
Viancour et al. 1988 61
Vining and Drewes 1985a, b 27
vitronectin 100
Vyklicky and Nicholls 1988 127

Wall and Taghert 1991 125, 154
Walters 1987a, b 131
Walters 1991 131
Walters and Ambron 1995 41, 42, 54, 56, 59
Walters et al. 1991 53
Walthall and Chalfie 1988 23
Walthall et al. 1993 23, 153
Wang et al. 1992 108
Webster and Hamilton 1972 16
Weeks 1990 3
Weeks and Ernst-Utzschneider 1989 32
Weeks and Truman 1986 32, 119
Weeks et al. 1992 32, 119
Wehrle and Chiquet 1990 103
Weiser et al. 1993 58
Westfall et al. 1994 14
Westin and Camhi 1975 152
White et al. 1986 23
White et al. 1991 9
White et al. 1992 23, 125, 153
Whitington 1979 148
Whitington 1993 98, 99
Wilkins 1994 140
Wilson et al. 1991 121
Wittenberg et al. 1990 97
Wojtowicz et al. 1989, 1991 161
Wong et al. 1981 113, 115, 121
Worden et al. 1988 148
Wu 1988 78, 89, 119
Wu et al. 1983 89

Xenopus 78, 80, 89
Xie and Barrett 1991 48
Xue and Grossfeld 1993 62

Yaross et al. 1986 15
Yawo and Kuno 1983, 1985 42
Yawo and Kuno 1985 44, 46, 49
Yogev et al. 1991 66
Young 1932 34, 56
Young 1972 152
Young 1973 152

Zhang and Nicholls 1983 99, 144, 153
Zhu et al. 1994 146, 148, 153
Ziller-Sengel 1965 20
Zimmerman 1986 66
Zipser and Cole 1991 107, 108
Zipser et al. 1989 106, 108
Zoran and Drewes 1987 27
Zoran et al. 1990 155
Zoran et al. 1993 156

Springer-Verlag and the Environment

We at Springer-Verlag firmly believe that an international science publisher has a special obligation to the environment, and our corporate policies consistently reflect this conviction.

We also expect our business partners – paper mills, printers, packaging manufacturers, etc. – to commit themselves to using environmentally friendly materials and production processes.

The paper in this book is made from low- or no-chlorine pulp and is acid free, in conformance with international standards for paper permanency.

MIX
Papier aus verantwortungsvollen Quellen
Paper from responsible sources
FSC® C105338

If you have any concerns about our products,
you can contact us on
ProductSafety@springernature.com

In case Publisher is established outside the EU,
the EU authorized representative is:
Springer Nature Customer Service Center GmbH
Europaplatz 3, 69115 Heidelberg, Germany

Printed by Libri Plureos GmbH
in Hamburg, Germany